Designing Transportation Systems for Older Adults

Human Factors and Aging Series

Wendy A. Rogers
University of Illinois Urbana-Champaign

Given the worldwide aging of the population, there is a tremendous increase in system, environment, and product designs targeted to the older population. The purpose of this series is to provide focused volumes on different topics of human factors/ergonomics as they affect design for older adults. The books will be translational in nature, meaning that they will be accessible to a broad audience of readers. The target audience includes human factors/ergonomics specialists, gerontologists, psychologists, health-related practitioners, as well as industrial designers. The unifying theme of the books will be the relevance and contributors of the field of human factors to design for an aging population.

Designing Transportation Systems for Older Adults
Carryl L. Baldwin, Bridget A. Lewis, Pamela M. Greenwood

Designing for Older Adults
Principles and Creative Human Factors Approaches, Second Edition
Arthur D. Fisk, Sara J. Czaja, Wendy A. Rogers, Neil Charness, Sara J. Czaja, Joseph Sharit

Designing Displays for Older Adults
Richard Pak, Anne McLaughlin

Designing Telehealth for an Aging Population
A Human Factors Perspective
Neil Charness, George Demiris, Elizabeth Krupinski

Designing Training and Instructional Programs for Older Adults
Sara J. Czaja, Joseph Sharit

Designing Technology Training for Older Adults in Continuing Care Retirement Communities
Shelia R. Cotten, Elizabeth A. Yost, Ronald W. Berkowsky, Vicki Winstead, William A. Anderson

For more information, please visit: www.crcpress.com/Human-Factors-and-Aging-Series/book-series/CRCHUMFACAGI

Designing Transportation Systems for Older Adults

Carryl L. Baldwin
Bridget A. Lewis
Pamela M. Greenwood

CRC Press
Taylor & Francis Group
Boca Raton London New York

CRC Press is an imprint of the
Taylor & Francis Group, an **informa** business

CRC Press
Taylor & Francis Group
6000 Broken Sound Parkway NW, Suite 300
Boca Raton, FL 33487-2742

International Standard Book Number-13: 978-1-4822-4471-7 (Paperback)
978-0-367-25540-4 (Hardback)

Library of Congress Cataloging-in-Publication Data

Names: Baldwin, Carryl L., author. | Lewis, Bridget A., author. | Greenwood, Pamela M., author.
Title: Designing transportation systems for older adults / by Carryl L. Baldwin, Bridget A. Lewis, and Pamela M. Greenwood.
Description: Boca Raton : Taylor & Francis, a CRC title, part of the Taylor & Francis imprint, a member of the Taylor & Francis Group, the academic division of T&F Informa, plc, 2019. | Series: Human factors and aging series | Includes bibliographical references.
Identifiers: LCCN 2019006049| ISBN 9781482244717 (paperback : alk. paper) | ISBN 9780367255404 (hardback : alk. paper) | ISBN 9780429162336 (ebook)
Subjects: LCSH: Older people--Transportation. | Roads--Design and construction. | Roads--Safety measures. | Pedestrian facilities design. | Autonomous vehicles.
Classification: LCC HQ1063.5 .B35 2019 | DDC 305.26--dc23
LC record available at https://lccn.loc.gov/2019006049

Visit the Taylor & Francis Web site at
http://www.taylorandfrancis.com

and the CRC Press Web site at
http://www.crcpress.com

For Raja,
… who contributed much to our understanding of people over
the age of 65, yet sadly did not get to experience it himself.

Contents

Preface

Throughout this book we have sought to bring together for the first time comprehensive information needed to assist with all aspects of designing, delivering, and evaluating transportation systems for use by older adults. We further present the necessary supporting background on aging and human factors issues as well as practical guidelines needed to accommodate older adult transport users.

This book aims to provide up-to-date, practical advice based on the most recent scientific knowledge and best practices on how to design transportation systems and services to meet the needs and the sensory, cognitive, and physical capabilities of older adults. Focus is placed on informing practitioners, system designers, and policy makers about age-related changes that can affect interactions with transportation systems. It presents best practices, design guidelines, and methods of accommodating the needs of older adults while improving safety and mobility for all.

Although there are a few existing books on issues related to human factors and transportation and even a few specifically on older drivers, none of the existing books place emphasis on the mobility needs of older adults across the wide range of transportation modalities currently available, from walking to air travel. Older adults are the fastest-growing segment of the driving population and are increasingly maintaining active lifestyles that rely on many forms of transportation. Physically active lifestyles are increasingly recognized as very important for maintaining cognitive and brain health late in life. Moreover, negative effects of social isolation on mortality make it important that older people retain their mobility. Further, transportation systems are becoming increasingly complex as new technologies emerge. It is essential that this fast-growing segment of the population be considered in the design process. This book brings together the information necessary for this to occur in an easily accessible and comprehensive format.

As much as possible, we tried to present clear design guidance aimed at improving transportation usability among older adults. We bring our collaborative expertise in the field of human factors and transportation, and knowledge of aging and age-related transportation needs and user capabilities to this work. We cover a wide range of transportation systems, including the notably important issue of older drivers; but we also consider additional transportation forms, including public transportation via bus and subway, air transport, rail, and even bicycling and walking.

Throughout we make use of personal vignettes and numerous examples of best practices based on both the scientific literature and our collective expertise. Each chapter contains summaries, usable guidelines, and resources for further reading on specific topics. Although our recommendations are based on peer-reviewed literature, we made a strong effort to avoid scientific and academic jargon and provide information in a nontechnical fashion understandable by the practitioner community.

We provide useful background about normative age-related changes in abilities and how these changes impact the ways older adults interact with transportation systems. We believe that the use of this information has the potential to greatly improve the safety and mobility of older adults and can be used to inform policy, community health, and safety practices as well as to improve market penetrability by product manufacturers.

Authors

Carryl L. Baldwin earned her PhD in human factors psychology from the University of South Dakota in Vermillion, South Dakota, in 1997. At the time of this writing, Dr. Baldwin was an Associate Professor and Director of the Human Factors and Applied Cognition Program at George Mason University in Fairfax, Virginia. In Fall 2019 Dr. Baldwin became the Carl and Rozina Cassat Distinguished Professor of Aging and Director of the Regional Institute of Aging at Wichita State University in Wichita, Kansas. Her primary research interests are in the areas of auditory and multimodal display design, alarms, advanced driver assistance systems (ADASs), driver behavior, mental workload assessment, aging, operator state classification, and human-automation interaction. Her previous publications include *Auditory Cognition and Human Performance* (2012, CRC Press) as well as numerous scientific journal articles, book chapters, and conference proceedings. Dr. Baldwin also has expertise in issues pertaining to attention management in autonomous systems, neuroergonomics, and the driving behavior of high crash-risk populations including older adults and fatigued and distracted drivers.

Bridget A. Lewis earned her PhD in human factors and applied cognition from George Mason University in 2017. Dr. Lewis is currently working as a Human Factors Engineer for the MITRE Corporation in McLean, Virginia. Her research interests include multimodal display design, advanced driver assistance systems, medical human factors, aviation systems, and enhancing accessibility for transportation systems. Dr. Lewis's previous publications include scientific journal articles, conference proceedings, and book reviews. The author's affiliation with the MITRE Corporation is provided for identification purposes only and is not intended to convey or imply MITRE's concurrence with, or support for, the positions, opinions, or viewpoints expressed by the author.

Pamela M. Greenwood earned her PhD in physiological psychology from the State University of New York (SUNY) at Stony Brook in 1977. Dr. Greenwood is currently an Associate Professor of Psychology at

George Mason University in Fairfax, Virginia. She has long-standing research interests in cognitive aging, cognitive training, and the genetics of cognitive aging in both healthy aging and Alzheimer's disease. Her previous publications include *Nurturing the Older Brain and Mind* (2012, MIT Press) and many other peer-reviewed journal articles and book chapters. Dr. Greenwood has expertise in cognitive genetics, use of transcranial direct current stimulation in cognitive training, and effects of cognitive aging and Alzheimer's disease on attention.

chapter one

Introduction and scope

> A 91-year-old motorist ran over two people Monday morning while backing out of a parking space at the Mar Vista Post Office parking lot off Grandview Boulevard, according to Los Angeles police. A 74-year-old woman struck by the SUV died at the scene and an 83-year-old man, believed to be the victim's husband, was taken to a local hospital with serious injuries.
>
> *Argonaut Online,* December 15, 2014

The incident occurred in broad daylight, at low speed, and yet tragically took the life of an unsuspecting pedestrian. Accounts like these are all too frequently in the news, even if they are not that common relative to the number of miles older adults drive. They catch the attention of media. Sometimes such incidents occur while the older driver is backing up, unable to see what is behind him or her. Other times they are due to the driver inadvertently hitting the gas pedal instead of the brake.

At the same time, there are news reports of older people voicing sentiments such as this: "'You know what I'd do if I couldn't drive?' an older man barks from inside his home, with a nod to a living room recliner. 'I'd go sit down right there in that chair and die...'" This quote was from an 80-year-old Detroit driver who was interviewed for a story about older drivers that aired on WZZM Channel 13 ABC on May 8, 2014.

Designing safe and effective means of meeting the mobility needs of older adults is a global imperative. For many companies it is not only the right thing to do, it will also be financially beneficial. Human factors design solutions for the personal automobile are one avenue for achieving this imperative by assisting the older driver. In general, the term *older* refers to anyone over the age of 65 years. However, the tremendous variability in the aging process means that any two individuals over age 65 are likely to be different in more ways than any two 20-year-olds. Despite the increased variability associated with age, general changes in sensory, perceptual, cognitive, and physical abilities tend to accompany advanced age and can lead to increased difficulty interacting with transportation interfaces.

Drivers over age 80 years are involved in more fatal crashes and are more likely to be at fault in fatal crashes than any other age group.

Older pedestrians are more likely to be struck when crossing streets and navigating parking lots than their younger counterparts. However, the great variability among older adults means that strict cutoffs for things like driver's licensure are both unethical and ineffective. Any given octogenarian driver may be safer behind the wheel than a random 46-year-old, and likely to be safer than the average 20-year-old male driver. The National Highway Traffic Safety Administration (NHTSA) currently focuses on what is known as the three D's of traffic collisions (drunk, drugged, and dangerous). Young males are much more likely to engage in dangerous driving behaviors like speeding and passing in narrow gaps than their older counterparts. Providing in-vehicle adaptive speed control devices may assist older adults by off-loading part of the control aspect of the driving task while assisting young drivers with maintaining safe and legal speeds. In fact, we often see that design recommendations that improve safety for older adults improve safety for all. The few exceptions to these design guidelines are pointed out in subsequent chapters.

Further, improved design in domains such as roadway infrastructure and driver vehicle interfaces (DVIs) can have dramatic, immediate impact that is difficult to achieve through training and attempts to change well-established cultural safety norms among the millions of people involved. Technological advances are making driving and flying safer for all ages. As we discuss in the current volume, there are many design factors that can make transportation in all forms safer for people of all ages, and for older adults in particular.

Population aging is a well-known reality in the United States and Europe. This demographic trend of population aging is expected to continue (Figure 1.1). The group referred to as the baby boomers (those born after World War II between 1946 and 1964) are reaching retirement age, but many are choosing not to retire for a variety of personal and economic reasons or are choosing to have "encore" or second careers after retiring from their first career. Today's older adults are expecting to have more mobile lifestyles than their predecessors, regardless of whether or not they remain employed.

Data from the U.S. Census confirm this population aging trend by indicating that not only are the percentages of adults over 65 years of age increasing, people over the age of 85 are the fastest-growing segment of the population (Figure 1.2).

Older adults today are more active, healthy, and technologically savvy than those of previous generations, and they live more mobile lifestyles. They rely on many modes of transportation to meet their needs. However, personal automobiles remain the most prevalent mode. The most recent data compiled by the U.S. Department of Transportation (2017) in its National Household Travel Survey indicate that 90% of all trips are taken by personal vehicle across all age groups over age 65 years. Designing

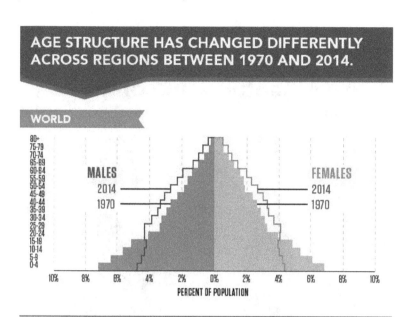

Figure 1.1 Population aging trends based on UN figures. (Used with permission from the Population Reference Bureau, Washington, DC, https://www.prb.org)

transportation systems that effectively meet the needs of this fast-growing segment of the population is a global imperative. Providing design guidelines to accomplish this imperative is the aim of this book.

To illustrate the issues to be discussed in subsequent chapters, we begin by presenting two fictional personas in specific situations that highlight some of the strengths as well as some of the challenges and issues faced by older adults. Personas are archetypal or prototypical users and are frequently used to aid design as they facilitate a user-centered design approach. Later in the text, we refer back to these personas, as well as introduce new ones, and discuss lessons learned from working with older adults in person. These personas can illustrate how key design recommendations can be applied in a wide variety of transportation modes ranging from the personal automobile to air travel and other forms of public transportation.

Before we can improve transportation systems for older adults, we need to understand more about their general mobility, driving habits, and preferences for modes of transportation. In the remainder of this chapter we discuss important general characteristics of today's older adults, how this demographic is changing, and how they choose to meet their transportation needs.

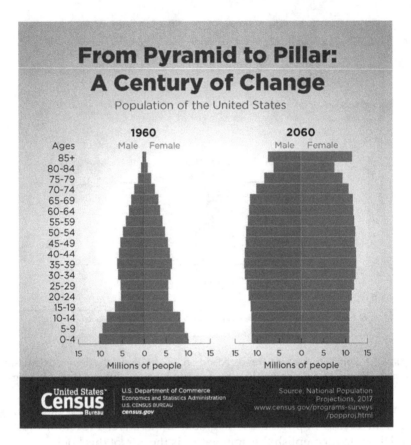

Figure 1.2 Percentage of U.S. adults over the age of 65 and 85 years.

1.1 Population aging

Declining fertility rates and increasing life expectancies around the globe in many countries are resulting in a phenomenon that has come to be called *global population aging*. That term means that the median age of a population is rising. Populations in the United States, Europe, Australia, Asia, South America, and most other countries are aging. This trend has numerous ramifications for designers of transportation systems. A larger proportion of the consumer market will soon be over 65 years of age. Older adults will be a larger share of the drivers and purchasers of vehicles, airline tickets, bus passes, etc. Therefore, in addition to being a global imperative to maintain the safety of this growing proportion of the population, it will be financially prudent for the companies and suppliers of transportation systems to do so. It is worth getting to know more about the lifestyles and habits of this changing demographic.

1.2 Mobile but less active lifestyles

Older adults from most, if not all, developed countries are increasingly sedentary, and this less active lifestyle is accompanied by a greater risk of cardiovascular disease related to obesity, hypertension, and diabetes. These trends are particularly evident in North America, where over half of the U.S. population is currently considered overweight or obese. Only recently has the difficulty of losing weight been recognized, with only 15%–20% of obese people being successful in permanent weight loss. In light of the serious health consequences of obesity (e.g., diabetes, heart disease, sleep apnea, and dementia), there is a public health motivation to make it easier and safer for people to walk and bicycle safely.

At the same time, these older adults in the United States retire later than their predecessors. Further, a substantial number of people who do retire subsequently reenter the workforce within the next 2 years. According to a recent AARP study (Brown, 2003), only 40% of workers between the ages of 50 and 70 expect to quit working before age 70, and over 25% expect to keep working until age 80 or beyond. These workers will continue to rely on transportation services to get to and from work on a daily basis and will travel using a variety of means for both work and pleasure.

1.3 Lifestyle factors

Older adults generally have poorer health, less money, but more time than younger adults. Increasing numbers of adults in Western European countries and the United States live in single households. Today's older adults are also more technologically savvy and less likely to see themselves as having any type of sensory, cognitive, or physical limitations relative to previous generations. This lack of insight into their actual abilities can sometimes create problems for older people who may attempt travel without being aware of the accommodations that their physical and cognitive limitations may require.

Today's older adults, unlike previous generations, are also more likely to be living farther away from family and other support networks and thus have an even stronger desire for independent transportation. More than before, octogenarians are likely to travel to see their children and grandchildren and for entertainment and leisure.

1.4 Importance of mobility

Mobility in older people is very important for health, social inclusion, and quality of life. According to a 2010 U.S. Centers for Disease Control and Prevention (CDC) Report on Healthy Aging, impaired mobility is associated with a number of adverse outcomes including such health problems as cardiovascular disease and depression. A number of large prospective studies in Europe and the United States have demonstrated that social isolation (based on contact with friends, family, and civic participation) increases all-cause mortality with a magnitude similar to widely recognized risks such as smoking and alcohol consumption. Underscoring the importance of this global issue, the CDC's Director of Unintentional Injuries Division, Grant Baldwin, spoke before the Special Committee on Aging of the U.S. Senate to discuss the importance of mobility for the health and safety of older adults, in particular. A portion of Baldwin's address spoke specifically to the importance of mobility:

> Mobility – whether by car, foot, bicycle, public transit, or other transportation options, such as ride sharing, shuttles, or volunteer driver services – enables older adults to remain independent, active and socially connected. Mobility concurrently helps older adults obtain needed health care and preventive care services, and access other health-promoting goods and services. Ease of mobility also may enable older adults to pursue volunteer or paid work opportunities, further connecting them to communities as well.
>
> **(Baldwin, 2010, pp. 3–4)**

> In addition, the opportunity to walk to a destination, or combining walking with another form of transportation – like public transit – enables, facilitates, and encourages older adults to be physically active, thus reducing their risk for obesity, diabetes, heart disease, stroke, depression and other chronic health conditions. Communities need to be safe for older adults and all pedestrians to walk throughout the community, and the CDC supports evidence-based interventions that encourage healthy lifestyles while promoting safety.
>
> **(Baldwin, 2010)**

In 1997, the United Kingdom government issued a paper discussing the importance of integrated transportation to the goal of making a fairer and more inclusive society (see discussion in Church et al., 2000). Subsequently, an international nonprofit organization referred to at the time as "Help the Aged" set up a Transport Council to address the needs of older adults. Help the Aged had noted a lack of adequate consideration of the mobility needs of older adults in the broader plan. The resulting Transport Council, made up of bus and train companies, users, planners, academics, and professionals, solicited input from older adults across the United Kingdom about the mobility issues they faced. Their subsequent report (Transport Council, 1998) identified safety, accessibility, reliability, and affordability (SARA) as the primary issues to be addressed to prevent loss of mobility and social isolation among older adults.

In sum, limitations in mobility and unfulfilled desire for travel can negatively impact older adults' quality of life. Numerous socioeconomic, health, financial, and other demographic factors impact older persons' interactions with transportation systems, which, in turn, has personal and social costs.

1.5 Modes of transportation

The personal automobile is the major form of transportation for older adults living in the United States, the United Kingdom, and other industrialized countries. Access to public transportation, like buses and trains, varies widely by geographical location. Many older adults in the United States live in suburban or rural areas where public transportation is limited at best. However, even when access to buses and other forms of public transport is available, older adults prefer the convenience of the personal automobile. Ride-hailing services are a solution that is well-suited for older people living in urban and suburban areas, though less so for those in rural areas. However, ride-hailing services typically require smartphones. According to a February 2018 PEW Research Center report (http://www.pewinternet.org/fact-sheet/mobile/), although approximately 85% of adults over age 65 have a mobile phone, only about 46% of those mobile phones are smartphones. There are efforts to circumvent the need for smartphones to use ride-hailing services (e.g., "gogograndparent.com"). Training aimed at helping older people use public transit has the potential to somewhat offset their preference for personal cars, if designed effectively. This issue is discussed in Chapter 8. But, DVIs that match the needs and capabilities of older adults remains critical. We discuss effective DVI design for older adults in Chapter 4. Effective DVI design is a critical component of maintaining mobility because the personal automobile remains the transportation mode of choice for most older adults.

1.6 More drivers driving more miles

The number of drivers over the age of 65 rose dramatically from the 1970s to the 1990s, with the number of older females obtaining licensure increasing in particular. For one, population growth during the baby boomer generation has resulted in a greater number of adults reaching old age than in previous generations. This general increase along with medical and technological advances, and the trend to move to suburban areas where public transportation services are often not financially feasible (e.g., low ridership and lack of existing infrastructure dissuade local governments from investing the funds) have led to more older adults obtaining and retaining their drivers' licenses and driving more miles than their predecessors in previous generations.

1.7 A global priority

Population aging is a global phenomenon that is particularly widespread in developed countries but also present in developing countries. It is estimated that two-thirds of all people who have reached 65 years of age are still alive. Global population aging is a trend that has the potential to transform societies, demanding shifts in financial expenditures and restructuring of the family, the labor force, and politics, to name a few. Safe, accessible, reliable, and affordable transportation is essential to allow the growing older segment of the population to remain independent yet socially connected, maintaining a high quality of life. Incorporating designs that accommodate the needs and capabilities of older adults in all forms of transportation has the potential to dramatically improve the safety and mobility of older adults. Further, designing transportation systems for older adults has the potential to facilitate both a higher-quality lifestyle for a large segment of the population as well as greater market penetration for new products, as this segment of the population gains consumer power.

Because a majority of older adults in the United States rely on the use of a personal automobile to meet most of their transportation needs, we end this chapter with two fictional persona depictions of older drivers. These personas illustrate key issues facing older drivers.

1.8 Personas of older adults

1.8.1 Joe Green using a navigation system while on vacation

Joe Green, a 73-year-old professor, is returning from a conference trip with his wife, Sally. Joe is active in his community and has an extensive social life. Generally he does most of the driving for himself and Sally. He loves to travel and is generally the one to make nearly all the travel arrangements for the frequent trips he takes. To date he has not noticed any trouble with his driving, nor does he ever have trouble getting around in his familiar neighborhood. He drives to work every day, though he and Sally generally prefer to take the subway into the city, rather than drive from their suburban home. Today is a different story. Joe is now trying to negotiate his way through an unfamiliar city in a rental car to catch a flight. The road signs seem confusing and difficult to read. There are pavement markings that he has never seen before, and he is not sure what they mean. It has just started raining, and he attempts to turn on his windshield wipers only to realize again that he is unfamiliar with the controls in this rental vehicle. "How do you turn on the windshield wipers?" he asks out loud as he struggles to understand the unfamiliar controls. He feels himself getting more and more frustrated and stressed as he searches the dashboard for some indication of what time it is. They are running behind schedule due to difficulty checking out of their hotel. He wonders if they will make their flight if he misses his next exit. Realizing that he is exceeding the speed limit for this roadway, Sally tells him to slow down just as the rental car navigation system announces their next turn. Joe cannot quite hear it. He thinks it said, "Turn left on highway 84," but they have now just passed it, and he is still trying to turn on the windshield wipers. He glances at the navigation systems' visual display but cannot make out clearly the words on the screen. He thinks to himself, "How can they get the navigational system to repeat the instruction and to present the information louder so that I can clearly hear it?" Sally tries to help him figure out how to read the visual display to clarify the instruction, while he finally finds the switch for the windshield wipers and looks for a way to reroute in case they have missed their exit. Sally is also having difficulty reading the display. How could this navigation system be more user friendly? Could there be more easily accessible options for increasing the volume and command distance to facilitate use by a wider range of users, including older adults? Frustrated, a bit frazzled, and short on time, Joe and Sally make it to the rental car drop-off location where they are then met with another confusing array of signs. They struggle to get their luggage onto the shuttle bus and then arrive at the airport where they must now figure out their gate and arrive there before boarding is closed. How could there be a way to gather the information they need while on the shuttle bus and perhaps even request assistance en route?

Joe's scenario highlights some of the many mobility issues that become more challenging with age. Driving involves operational, strategic, and tactical skills. Operational skills, such as maintaining control of the vehicle and reacting appropriately to changes in the traffic patterns, are the most fundamental aspect of driving. These are frequently handled with ease when one is driving a familiar vehicle under normal driving conditions. But, Joe is driving a rental car and is therefore unfamiliar with many of the controls, including the windshield wiper interface. Tactical skills such as maintaining a safe speed, navigating lane position, and obeying the rules of the road are also important. Tactical skills can generally be performed well by experienced drivers under normal conditions. But, bad weather due to rain or fog can reduce visibility, making traffic signs more difficult to see, particularly for older adults who may already be experiencing the challenges of impaired visual acuity and a narrowed field of view. The increased stress from being late and the increased cognitive demand of strategic skills such as navigating in an unfamiliar area result in Joe not noticing the speed limit sign, having difficulty with his tactical skills, and being unable to follow directions from the unfamiliar pavement markings.

This example also illustrates common design issues with in-car navigation systems. Their visual displays are often difficult for older adults to see due to multiple issues, such as small font size, poor contrast displays, and glare. Voice guidance can assist drivers, but the voice guidance must be clearly audible above background noises. As discussed in Chapter 3 aging is accompanied by hearing loss; therefore, voice guidance needs to be sufficiently loud for ease of processing. Older adults take longer to switch their attention from one visual target to another and have generally slower processing time. Therefore, guidance systems need to provide sufficient time for drivers to maneuver before turns and should provide an easy means for drivers to request repeat instructions if key guidance commands are missed.

The confusing signage at the airport and physical challenges of dealing with their luggage are all too common. Chapter 10 addresses design recommendations for improving the comfort and safety of older airline passengers.

1.8.2 Gertrude the older rural driver

Gertrude is in her early eighties; she is a retired school bus driver and homemaker. She is in pretty good health, though her knees tend to hurt considerably, particularly if she has to walk up or down stairs or inclines. Gertrude lives on the family farm in rural Ohio where she is the primary caretaker for her husband. She does all of the driving since her husband's stroke 10 years ago. There are not many transportation alternatives where she lives, other than calling a friend or relative for a ride, something she rarely does as she is proud of her independence. However, she has changed some of her driving habits. For instance, she tries not to drive long distances, and she avoids driving at night, into the setting or rising sun, and in big cities, if she can. She knows that at some point she may have to give up driving and that will likely involve moving to town. But until that time, she prefers to live in the country on the family farm, which has been her home for over 40 years. Winters are the worst. The roads can be pretty rough when it snows and can be completely blocked if the wind comes up. She has some issues with arthritis and has limited agility. For example, she finds it difficult to turn her head to see something behind her while driving and would not welcome a trudge through the snow even for a short distance if she were to get her vehicle stuck. Easy access to in-vehicle weather, road reports, and mobile communications and automated emergency assistance could be invaluable to Gertrude.

Figure 1.3 Image of Apple CarPlay appearing on a vehicle console screen. The advantage is that navigation maps visible on the driver's mobile phone are also displayed in a larger format on the console screen. (Personal image of the author, Pamela Greenwood.)

Gertrude could really benefit from a well-designed communications interface. She is not willing to use a complicated system that will take considerable time to learn. However, a clear, simple interface such as Apple's CarPlay iPhone (shown in Figure 1.3) or Android Auto interfaces would allow her to easily call for help if she found herself stranded on the road or stuck in snow. Although she is still learning to use her smartphone, it would also assist her when navigating through unfamiliar areas and could provide driving times to places so that she could plan her trips to avoid challenging driving situations (i.e., night and rain). In Chapter 4 we discuss smartphone use among older people, and in Chapters 6 and 11 we discuss DVI design for current and advanced systems, respectively.

1.9 Overview of this book

The personas described illustrate some of the many design issues faced by older users of transportation systems. The aim of this book is to provide a concise guide to designing effective, efficient, and safe transportation interfaces and services to meet the needs and capabilities of older adults. Chapter 2 addresses the importance of mobility for maintaining a high quality of life and preventing a growing segment of our global population— older adults—from becoming socially isolated and dependent. Chapter 3 addresses age-related changes in sensory, cognitive, and physical abilities that have a direct impact on designing older-adult-friendly transportation systems. For example, we discuss the way the sensory and cognitive capabilities of drivers like Joe and Gertrude may impact use of different modes of transportation. Joe has difficulty hearing but still prefers to rely on auditory navigation commands while driving because he finds it difficult to see the visual display.

Chapters 4 through 6 specifically address older drivers, focusing first on prevalence rates, challenges faced, and ways that older people self-regulate their driving. Chapter 5 presents specific design recommendations for transportation infrastructure, and Chapter 6 specifically addresses the design of DVIs to fit the needs and capabilities of older adults.

Beginning with Chapter 7, the emphasis shifts to outside the personal automobile. Chapter 7 addresses the particular needs and design recommendations for assisting older pedestrians and cyclists. For example, redesign of signage at crosswalks can dramatically improve the safety of older pedestrians given that they are more likely to misjudge how long it takes them to cross a street. Chapter 8 provides design recommendations for transport training aimed at older adults. Numerous older driving training programs currently exist, but recent successes have been achieved by also providing training in how to use existing public transportation services such as buses and trains. Design guidance for development and successful implementation of these programs is discussed. Chapter 9 addresses design strategies for alternative forms of transportation, such as public transport buses, trains, and senior transportation systems. Beyond training older adults to use public transportation services, ensuring that the displays and interfaces they encounter are effectively designed is critical for safe and efficient use. Chapter 10 provides design recommendations for older passengers of air travel. Chapter 11 discusses how changing transportation technologies are likely to impact older adults. Automobiles, as well as buses, trains, and planes are becoming increasingly automated. In the future older adults who are no longer physically capable of driving may be able to program their automated vehicle to take them to doctors' appointments and shopping. These new technologies promise to have profound impacts on the mobility of older adults. At the same time, humans are not evolving as fast as technology, and the needs, limitations, and capabilities of the user may play an even more essential role in the design of future technologies. The final chapter (Chapter 12) synthesizes the design recommendations aimed to accommodate age-related changes in perceptual-cognitive abilities and promote safe, accessible, reliable, and affordable transportation systems. Chapter 12 also identifies gaps in our current design knowledge.

Throughout the book, our goal is to provide the most current knowledge about the topics of discussion to provide a state-of-the-science and practice in these areas. To enhance readability of the text, we minimize references within the text but instead provide recommended readings for each chapter for those desiring additional in-depth information. Moreover, in the chapters, where appropriate, we provide a summary list of the practice or design recommendations at the end of each chapter.

Recommended readings

Coughlin, J. E., & D'Ambrosio, L. A. 2012. *Aging America and Transportation: Personal Choices and Public Policy.* New York, NY: Springer.

Holt-Lunstad, J., Smith, T. B., & Layton, J. B. 2010. Social relationships and mortality risk: A meta-analytic review (social relationships and mortality). *PLOS Medicine*, 7(7), e1000316. doi: 10.1371/journal.pmed.1000316

Vettori, S. 2010. *Ageing Populations and Changing Labour Markets: Social and Economic Impacts of the Demographic Time Bomb.* Burlington, VT: Ashgate.

chapter two

Mobility and quality of life

Adults today are living longer and healthier lives than any previous generation. They are also experiencing a number of other positive trends including more time for leisure activities, recreational travel, creative pursuits, volunteerism, and additional or encore careers. Some retirees take up competitive sports, as illustrated in Figure 2.1. Mobility plays a critical role in the quality of life of older adults.

Independence and choice are highly valued among all adults, and this does not change as one ages. Mobility is essential to independence, and limitations in mobility negatively impact physical, psychological, and social aspects of life. Designing for safe, affordable, reliable, and accessible (SARA) transportation requires effective design and planning at multiple system levels, ranging from the design of buildings, communities, and neighborhoods, to design of public and private transportation. It is relatively well known that decrements in cognitive and physical capabilities are generally associated with decreased mobility. Less well known is the negative impact that reduced mobility can have on cognitive and physical capabilities. In this chapter, we discuss many of the mobility needs and challenges for older adults, the importance of mobility on quality of life, and ways of improving design to facilitate mobility. But first, we define what the construct *quality of life* means.

2.1 Defining quality of life

Quality of life is a multidimensional construct with an evolving and sometimes controversial definition. It has been defined indirectly as access to resources based primarily on objective economic factors. More recently it has been defined more directly by indices of subjective and psychosocial well-being. A third view integrates the previous two pointing to the fact that there is often only modest or no correlation between the two. At present there is no universally accepted definition of quality of life.

Despite the lack of a standard definition, there is recognition that quality of life contains subjective components and has distinctive characteristics at different stages of the life cycle. Modern views indicate that it consists of personal/social factors as well as subjective/objective factors. It is important to note that for older adults, health and social relationships are particularly important dimensions. There is much we can do to design communities

Figure 2.1 Former U.S. Olympic coach Payton Jordan of California sets a world record in the M80 age group in the 200-meter dash. (This file is licensed under the Creative Commons Attribution-Share Alike 3.0 Unported license. Subject to disclaimers.)

that support both mobility and social interaction. For example, the creation of public spaces with benches and places to gather encourages people to be more mobile within their neighborhoods and support both low-impact exercise by walking and socialization. Both of these activities have been shown to improve quality of life.

Data shed light on how older people view quality of life. Elosua (2011) examined responses to a paired comparison survey where over 300 individuals over the age of 65 years (with a mean age of approximately 71 years) were asked to rate which of two dimensions (from a total of five) were most important to a good quality of life. Participants were all community-dwelling individuals who lived in their homes and maintained relatively active lifestyles. The five dimensions were taken from those established in previous work (Farquhar, 1995). The dimensions were related to health, functional autonomy, social and family networks, home and environmental conditions, and social activities. Participants were asked to choose which of two statements was most relevant to a good quality of life, and statements consisted of items such as, "Having good health," "Having social and group activities," and "Having personal autonomy."

The most important dimension was health, with personal autonomy chosen as the second most important dimension. Neither gender nor age affected the perceived importance of the dimensions within the sample. Having an adequately adapted home (with grab bars in shower and tub, no trip hazards, etc.) and social/family support tied for third place, with maintenance of social activities rated as the least important of the five. Elosua (2011) noted that this ordering is in contrast to several previous studies that have used other methods and found that health and social contacts are generally listed as the two most important dimensions.

An open-ended question technique asking elders to list what they considered to constitute a quality of life led to eight categories, with social relations being the most frequently reported dimension (Wilhelmson et al., 2005). Other dimensions included health, activities, functional ability, well-being, personal beliefs and attitudes, their own home, and personal finances. Maintaining independence, social relationships, and access to health care and favored leisure activities are interrelated with mobility, and all play an important role in maintaining a good quality of life. Before discussing in more depth the leisure activities of older adults, a definition of mobility is provided.

2.2 Defining mobility

Mobility can be defined in a number of ways. A working definition should include aspects such as access to people, places, and things that an individual enjoys. It may also involve travel for leisure and recreation, maintenance of social networks, and access to medical care. It can also involve the ability to simply walk outside and around one's neighborhood to experience fresh air, visit neighbors, and access everyday necessities. As discussed further in a later section, more seniors than ever before maintain their mobility through maintenance of a personal driver's license and personal automobile. However, seniors may also rely on relatives and friends to meet their mobility needs. Physical disabilities as well as declining cognitive and health status can combine to create mobility challenges for older adults, with a concomitant toll on access to leisure activities that are essential to the maintenance of a high quality of life. Older adults who require walkers or other assistive devices may have difficulty simply opening heavy outside doors in order to breathe fresh air. For these individuals, providing electronically assisted door openers can mean the difference between whether they are able to go outside without assistance for leisure or emergency egress.

2.3 Leisure activities in older age

Greater understanding of the importance of mobility can be gained by examining the ways older adults spend their time. Data from the American Time Use Survey (ATUS) examining the ways different age groups spend their time indicate that people over the age of 55 years spend an increasing amount of time engaged in leisure activities (Marcum, 2013). As illustrated in Table 2.1, by the age of 65 years, adults in America report spending over 25% of their time engaged in leisure activities.

The types of leisure activities engaged in by adults over the age of 75 are presented in order of prevalence in Table 2.2. Note that walking for recreation occurs much less often than more sedentary activities like watching television and reading. This is concerning in light of recent evidence that lifestyle, including obesity, physical inactivity, and low educational attainment, accounts for one-third of Alzheimer's dementia cases in the United States and Europe (Norton et al., 2014). Exercise and mentally stimulating activities are important for maintaining physical and cognitive health. Therefore, environments that support and promote these activities are essential to support public health.

The personal automobile is a primary form of transportation for the majority of adults in the United States. It is not surprising that the independence and mobility associated with being able to drive are highly valued among older adults. At the same time, age-related changes in sensory, cognitive, and physical capabilities result in many older adults being faced with the need to limit or cease driving altogether for their safety and the safety of others. These issues are discussed in more depth in Chapter 3. We now turn attention to a discussion of the consequences of driving cessation.

2.4 Driving cessation and maintaining social connections

Remember the words of the Detroit driver presented in the opening paragraphs of Chapter 1 who indicated he would prefer to just die if he had to give up his driver's license? He is not alone in this sentiment. The consequences of surrendering one's driver's license are many. First, there are the rather obvious issues surrounding the general inconvenience of not being able to get where one wants to go when one wants to get there. Second, there are a host of social, emotional, health, and independence issues.

Table 2.1 Activities engaged in by age with an emphasis on leisure

	Age group								
	(15–24)	(25–34)	(35–44)	(45–54)	(55–64)	(65–74)	(75+)	Z	Sig.[a]
Caregiving	0.0103	0.0262	0.0178	0.0077	0.0066	0.0050	0.0037	−1.4112	
Communication	0.0428	0.0370	0.0364	0.0348	0.0360	0.0394	0.0388	−8.3607	***
Eating	0.0335	0.0410	0.0419	0.0421	0.0456	0.0495	0.0515	34.3652	***
Education	0.0532	0.0076	0.0039	0.0028	0.0015	0.0011	0.0010	−26.4640	***
Household labor	0.0514	0.0919	0.1032	0.1029	0.1069	0.1084	0.0980	27.6381	***
Leisure	0.1730	0.1526	0.1538	0.1703	0.2058	0.2524	0.2788	66.4713	***
Personal care	0.0019	0.0030	0.0041	0.0068	0.0077	0.0095	0.0108	−3.0425	**
Private personal care	0.0272	0.0243	0.0250	0.0260	0.0264	0.0250	0.0260	6.7894	***
Sleeping	0.4757	0.4323	0.4176	0.4133	0.4193	0.4359	0.4494	−30.3395	***
Travel	0.0488	0.0516	0.0533	0.0502	0.0456	0.0386	0.0284	−11.8796	***
Volunteering	0.0031	0.0028	0.0047	0.0042	0.0037	0.0038	0.0042	−0.1794	
Waiting	0.0015	0.0017	0.0020	0.0019	0.0020	0.0022	0.0018	6.9577	***
Work production	0.0775	0.1279	0.1363	0.1369	0.0929	0.0292	0.0076	0.6337	

Source: Adapted with permission from Marcum, C. S. 2013. *Research on Aging*, 35(5), 612–640.

Note: Data come from the pooled 2003–2008 American Time Use Survey. Z-scores and significance stars based on Tobit regressions of each dependent variable on linear age variable.

[a] Significance codes: $Pr(>|z|) = $ ***0.001, **0.01, *0.05.

Table 2.2 The prevalence of different leisure time activies in order of prevalence (1 = most prevalent) reported by adults over the age of 75 years

1	Watching television and movies (not religious)
2	Reading for personal interest
3	Relaxing, thinking
4	Playing games
5	Attending religious services
6	Listening to the radio
7	Walking, for recreation
8	Participating in religious practices
9	Using computer for leisure activities
10	Attending or hosting parties/receptions/ceremonies

Source: Data from Marcum, C. S. 2013. *Research on Aging*, 35(5), 612–640.

The AAA foundation (see the suggested readings) recently compiled a report of the existing literature pertaining to the impact of driving cessation on older adults. Driving cessation is associated with declines in social, cognitive, and physical functions, to name a few, and an almost doubled risk of increased depressive symptoms. Relative to older people who are still driving, former drivers are less likely to be employed and demonstrate reduced participation in outside activities. The authors point out that many of the associated physical, cognitive, and health factors are likely mutually causative, meaning that they contribute to the reason for driving cessation and then are exacerbated by it.

Several researchers have systematically examined the consequences of driving cessation. A challenge to research in this area is teasing apart the relative cause and effect in this situation, given that many of the precipitant challenges (e.g., poor health) that may be exacerbated by driving cessation are also the cause of the cessation. Careful control of sociodemographic and health-related factors in longitudinal studies are needed to determine how much of the change is a result of driving cessation and how much is a function of some other adverse condition that would have impacted social activities regardless of driving status. Several studies that have controlled for these sociodemographic and health-related factors have shown decreases in important quality-of-life areas after driving cessation. For example, a review of the literature conducted by the AAA Foundation for Traffic Safety noted that in longitudinal studies, reduced physical functioning was strongly associated with driving cessation even after controlling for sociodemographic factors and baseline health. Out-of-home activity levels also decrease when older adults stop driving. There are a number of consequences to being housebound in that manner, including social isolation, reduced health, poorer nutrition, etc.

Data from participants aged 60 and older in the Baltimore Epidemiologic Catchment Area Study showed that driving cessation was associated with a reduced network of friends and that this association was not mediated by the ability to use public transportation. As discussed in more detail in Chapter 9, there are a number of challenges to using public transportation that older adults face. These challenges likely offset the potential benefit that this alternative form of transportation could provide. Overall, it appears that driving cessation in older people is associated with negative consequences.

Social integration, activity level, and health status are linked to well-being, which is in turn linked to driving status. While young adults realize that they have many forms of transportation available other than personal automobiles (e.g., walking, biking, and public transportation), older adults tend to be more reluctant to turn to these alternatives. This reluctance can stem from limited physical mobility. Many older adults find it hard to negotiate stairs and escalators leading to station platforms and bus seats. Some find it difficult to walk the distances required to get to bus stops and train platforms, and some find riding a bike challenging. Many older adults are concerned with being caught in adverse weather while waiting outdoors for trains and buses. Then, there are the rural elderly who may have few, if any, alternatives to driving since public transport in rural areas is virtually nonexistent.

Driving cessation is associated with a reduction in the frequency of social activities common among older adults. Specifically, after people stop driving they tend to reduce the frequency of engaging in activities such as going to a movie, restaurant, or sporting event; performing paid or unpaid community work; playing cards, games, or bingo; and attending religious or nonreligious organizational meetings. Even after accounting for sociodemographic factors such as age, gender, race, marital status, education, income, housing type, health-related factors, general cognitive status, and self-reported vision and hearing abilities, Marottoli and colleagues found that although there was a general decline in activity levels over time in aging, driving cessation was negatively associated with participation in social activities above and beyond the level that could be accounted for by general aging and the other factors. Individuals who stopped driving had declines in activity rates that were three times those of the average decline of the cohort of 1,316 participants. Living in an urban area, with its access to alternative forms of transportation and greater access to socialization without the need to drive can offset this decline to some extent. Considered together, this evidence shows that there are real costs when older people must stop driving.

2.5 *Quality of life and mobility in cities*

Cities and urban areas offer a range of both positive and negative aspects that impact quality of life for older people. On the positive side, there is greater access to cultural and educational events and a wider array of transportation options, including taxis and "ride-hailing" services, such as Uber and Lyft, as well as public transportation options. On the negative side, there tends to be more crime, overcrowding, pollution, and traffic congestion. Quality of life in cities depends on a balance between the good and bad aspects associated with urban life.

In towns and cities there is greater access to delivery services for a wide variety of goods, such as groceries and freshly prepared meals. This can decrease the need to travel outside the home but does little to maintain opportunities for socialization. Mobility in cities can be enhanced by well-designed transport systems, enhanced measures to maintain safety and security, and urban planning. Urban planning is particularly important. In the Netherlands where there are many protected bicycle lanes, 17% of people over 65 years of age bicycle every day, in comparison to 27% of people under 50. Bicycling in older people not only facilitates mobility, it can also facilitate social connectedness. It should be noted that use of public transportation among older adults in the United States does lag behind many other developed countries. This is likely due, at least in part, to the lack of alternatives to the personal automobile in many rural areas throughout the United States. In cities, well-designed urban areas can provide spaces for recreation and socialization and therefore have the potential to offset some of the decreases in quality-of-life factors associated with driving cessation. In rural areas, older people who cease driving are more vulnerable to social isolation, which carries negative health consequences (discussed briefly in Chapter 1).

2.6 Urban design

Neighborhoods, including front yards, sidewalks, parks, squares, pedestrian-only areas, and the primary streets surrounding homes and workplaces, can have dramatic impact on mobility and quality of life. Ensuring that there are attractive places to stop to rest and to just sit and relax while outside encourages people of all ages to spend time in public spaces. But, such places to sit can particularly benefit older adults. People are less reluctant to take a walk if they know there will be ample rest stops along the way. Plus, benches provide a place to spend time and engage in informal social interactions. Public buildings with heating and air conditioning, and bathrooms near public parks ensure that people can get out of the elements and attend to basic needs without long walks back home. Entryways into public buildings are more accessible to older adults and persons with disabilities if they have electronically assisted door openers.

China is one country that has done well in designing urban parks with recreational equipment, benches, and aesthetic beauty in the form of flowers, trees, and other ornamental designs. See Figure 2.2 for an example of a recreational area in a public commons area in China. It is quite common to see older adults gathering and exercising in the urban parks scattered

Figure 2.2 Common-area public park in the People's Republic of China.

throughout cities such as Beijing and Shanghai. Importantly, the maintenance of social connections and activities are not just crucial to mental health, they are also critical to overall health and are associated with mortality risk.

2.7 Health-care maintenance

Maintaining mobility also plays a key role in maintaining health. Transportation to and from doctors' appointments for routine checkups and minor treatment appointments are an important aspect of health care for older adults. These visits tend to be needed more often as people age. The need for more frequent visits to optometrists, audiologists, as well as other specialists and general practitioners are increasingly common with advanced age. Difficulty securing safe, affordable, reliable, and accessible transportation can lead to failure to receive adequate follow-up and treatment for nonessential, but important, health-care appointments.

When a person can no longer safely drive, and particularly if he or she no longer lives with someone who drives, it is essential that alternative forms of transportation are provided to meet health maintenance mobility needs. This problem illustrates what has been termed the *social determinants of health*—the conditions in which people live, work, play, worship, and grow old that influence health and quality of life. To help older people obtain transportation to medical facilities, doctors and medical facility providers could make available to their older patients access to information about alternative transportation formats (e.g., cards with transportation website addresses and brochures with brief descriptions of alternatives and where to find more information about each one). Recently, the Blue Cross Blue Shield Association announced a partnership with the ride-hailing service Lyft to arrange rides to medical appointments for their insured members. These issues are discussed in more detail in Chapter 9.

2.8 Independence

Mobility is essential to maintaining independence. A driver's license is viewed as a rite of passage into growing independence and early adulthood for many teens. Loss of this status (giving up one's license) can bring a sense of both personal and physical dependence. For many it symbolizes transition from the Third Age (post-retirement productivity) to the Fourth Age (declining function, increased depression, illness, and lower overall quality of life). Providing access to alternative transportation modes can offset some of this sense of loss.

2.9 Transportation needs and challenges

Community design can improve the safety of both older motorists and older pedestrians. For example, roadways surrounding large retail stores are especially dangerous for older motorists and pedestrians (Dumbaugh & Zhang, 2013). Roadways surrounding big-box stores tend to have faster speed limits and are associated with higher crash rates and fatalities or near fatalities among both older motorists and pedestrians (Dumbaugh & Zhang, 2013). Reducing the speed limit on these roadways and providing turn lanes and protected left turn signals would decrease crash risk for older drivers. Older pedestrians tend to underestimate the time it takes them to cross intersections and other roadways, and may think they are more visible to other motorists than they actually are. Specific design recommendations for each of these transport situations are provided in subsequent chapters.

2.10 Design recommendations

- Public policies are needed to ensure accessibility and mobility for older adults in order for them to maintain a high quality of life.
- Maintain green public spaces with ample seating to promote physical exercise and socialization opportunities.
- Install green areas providing shade and natural cooling (trees, vines) to combat urban heat islands in areas where heat exhaustion is a risk.
- Install electronic door opening devices to facilitate egress through heavy external doors.
- Municipalities could design cards and brochures that list alternatives to the personal automobile with links and phone numbers for how to obtain additional information. These could be provided to doctors, medical facilities, and community centers to distribute to their patients/clients.

Recommended readings

Ball, M. S. 2012. *Livable Communities for Aging Populations: Urban Design for Longevity.* Hoboken, NJ: John Wiley and Sons.

Brown, C. J., & Flood, K. L. 2013. Mobility limitation in the older patient: A clinical review. *JAMA,* 310(11), 1168–1177. doi: 10.1001/jama.2013.276566

Chihuri, S., Mielenz, T. J., DiMaggio, C. J., Betz, M. E., DiGuiseppi, C., Jones, V. C., & Li, G. July, 2015. *Driving Cessation and Health Outcomes in Older Adults: A LongROAD Study.* Washington, DC: AAA Foundation for Traffic Safety.

Daatland, S. O. 2005. Quality of life and ageing. In M. L. Johnson (Ed.), *The Cambridge Handbook of Age and Ageing* (Part 4, Section 12, pp. 371–377). Cambridge, UK: Cambridge University Press.

Hash, K. M., Jurkowski, E. T., & Krout, J. A. (Eds.). 2015. *Aging in Rural Places: Policies, Programs, and Professional Practice.* New York, NY: Springer.

Marottoli, R. A., Mendes De Leon, C. F., Glass, T. A., & Williams, C. S. 1997. Driving cessation and increased depressive symptoms: Prospective evidence from the New Haven EPESE. *Journal of the American Geriatrics Society,* 45(2), 202–206.

Marottoli, R. A., Mendes De Leon, C. F., Glass, T. A., Williams, C. S., Cooney, L. M., Jr., & Berkman, L. F. 2000. Consequences of driving cessation: Decreased out-of-home activity levels. *Journals of Gerontology: Series B: Psychological Sciences and Social Sciences,* 55B(6), S334–S340. doi: 10.1093/geronb/55.6.S334

Mezuk, B., & Rebok, G. W. 2008. Social integration and social support among older adults following driving cessation. *Journals of Gerontology Series B: Psychological Sciences and Social Sciences,* 63(5), S298–S303. doi: 10.1093/geronb/63.5.S298

chapter three

Sensory, cognitive, and physical challenges of aging specific to transportation

Aging is accompanied by sensory, cognitive, and physical changes that independently and interactively impact mobility. Because these changes occur gradually over time, they are not always obvious to the aging person but can have profound effects, nonetheless. Many age-related changes can be accommodated by the use of good design. In addition to the gradual nature of most age-related changes, there are large individual differences in the progression of those changes that preclude establishing strict age-based criteria. In this chapter we provide an overview of the changes that can be expected and discuss design recommendations that can facilitate safe mobility among older adults.

3.1 Peripheral changes

Peripheral changes are those age-related changes that affect sensory capabilities. Acuity changes occur in each of the major sensory systems, (auditory, visual, taste, smell, and tactile). Declining sensory abilities can pose problems, such as making it more difficult to read road signs, hear navigational directions, and feel the rumble strips at the side of the road. Declining sensory abilities can also exacerbate or be mistaken for cognitive changes. This distinction is important from a design perspective, because there are many ways of improving the salience of information displays and augmenting sensory processing that may greatly benefit older adults and even compensate for some cognitive decline. We begin by discussing age-related changes in hearing and vision, as these are the predominant sensory systems for obtaining information from the world.

3.1.1 Vision

The prevalence of visual impairment increases with age and poses many risks to safe mobility. Older drivers may have more difficulty seeing traffic signs and negotiating complex intersections. Older pedestrians may have more difficulty judging the speed of oncoming cars.

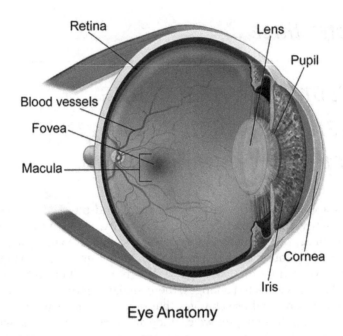

Eye Anatomy

Figure 3.1 Eye anatomy. (Adapted from Blausen.com staff. 2014. *Wiki Journal of Medicine*, 1[2]. doi: 10.15347/wjm/2014.010.)

Age-related visual impairments are more prevalent in women than men. Vision relies on a host of peripheral structures (e.g., cornea, pupil, and lens as illustrated in Figure 3.1), as well as central mechanisms, including the retina and visual cortex. Pinpointing the precise contribution of peripheral versus central mechanisms to age-related changes in vision has been an ongoing challenge for vision scientists.

3.1.2 Contrast sensitivity

Contrast sensitivity, or the ability to detect differences in luminance or color, is an important aspect of vision. It is particularly important for being able to detect objects in low light (e.g., nighttime) and in the presence of fog or glare. Contrast sensitivity declines with age and has been found to be a better predictor of the ability to read road signs than overall visual acuity (Evans & Ginsburg, 1985). Research has also found a relationship between contrast sensitivity and rapid deceleration events—a proposed surrogate for crashes (Chevalier et al., 2017).

Figure 3.2 illustrates the functional result of age-related changes in a number of visual processes, including loss of accommodation and contrast sensitivity.

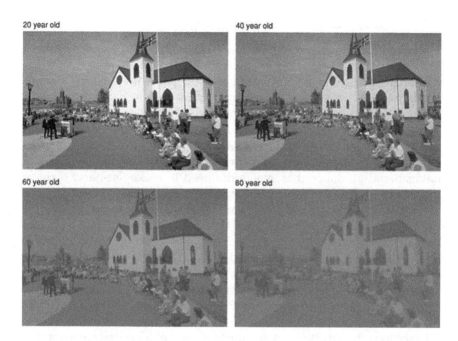

Figure 3.2 Changing view across the life span. (Borrowed from Margrain, T. H., & Boulton, M. 2005. In M. L. Johnson [Ed.], *The Cambridge Handbook of Age and Ageing* [p. 123]. Cambridge, UK: Cambridge University Press. With permission from Cambridge University Press.)

3.1.3 Temporal resolution

Aging is also accompanied by declines in temporal resolution, with older adults having difficulty detecting temporal changes and resolving spatial detail, particularly in low-contrast conditions. In other words, it takes longer for older adults to process rapidly changing stimuli. For example, an older adult can be expected to have more difficulty than a younger person detecting a blinking directional indicator or turn signal in bright light (a low-contrast situation in which the contrast between the blinking light and its surroundings is not that different). In current vehicle models, flashing rear lights are used to indicate that a car is turning (if only one side is flashing) or when both sides are flashing the presence of a hazard (such as a stopped vehicle). Older adults are more likely than young to have difficulty detecting the flash and distinguishing between the two types if the flash is detected. Slowing the flash rates to 12–20 per second and/or increasing the contrast and luminance of the flashing source will facilitate detection rates in older adults under most conditions.

3.1.4 *Motion processing and gap detection*

Visual motion processing declines with age. The older eye has reduced ability to detect motion and to discriminate the direction and speed of motion that is detected. Impaired motion perception poses obvious challenges to a wide range of mobility issues, such as judging the speed of oncoming traffic both as a pedestrian and a motorist. This is one of the reasons why older adults are more likely to have crashes while making left-hand turns at intersections without left turn lights. Providing left-hand turn signals (e.g., left-turn arrows) at busy intersections or navigational guidance that avoids the need to make left-hand turns at these nonsignalized lights can enhance the safety of older motorists.

Urban designers should be aware that older pedestrians frequently misjudge their ability to perceive the time needed to safely cross a street—that judgment must take into account both the speed of oncoming traffic and the time needed to traverse the distance. This results in the somewhat counterintuitive finding that marked crosswalks (pedestrian crossings) can actually decrease safety unless they are accompanied by traffic signals and signage alerting motorists to the presence of pedestrians. Marked crosswalks need to be designed to improve the visibility of pedestrians, and signalized intersections should allow ample time for slower older pedestrians to safely cross. This topic is discussed in more detail in Chapter 7.

3.1.5 *Glare and night vision*

Older adults often express difficulties with driving at night, particularly due to difficulties with glare. Though night-vision difficulties are a result of many factors, one key issue is that older eyes take longer to adjust to changing light levels. Greater susceptibility to glare also makes reading computer screens and electronic displays (e.g., train station departure tables) more challenging. Driving at sunrise or sunset poses a particular challenge for older drivers, since solar glare is more intense than headlight glare. Oncoming headlights and the glare of headlights in rearview mirrors during nighttime driving are another problematic source of glare and are frequently listed as a primary reason for self-restrictions in nighttime driving. Installing electrochromic or auto-dimming mirrors can reduce the nighttime glare from rearview mirrors.

The lens is the major focusing mechanism of the eye (see Figure 3.1). With age the lens becomes more rigid and bends less easily. This results in loss of accommodation, which makes it increasingly difficult to bend the lens enough to focus on things that are very close. Age-related loss of accommodation is called *presbyopia*, and it increases across the life span, noticeably affecting most people by middle age. The majority of people, starting generally in their mid-forties, require some form of reading glasses that aid in focusing on things up close. With advanced age, this difficulty in seeing up close, and seeing small fonts or detail in particular, increases. Simply increasing the font size on dials, labels, and displays can help offset this age-related loss of accommodation. Distance vision (myopia) changes little with aging unless cataracts are developing.

3.1.6 Color

Another common difficulty associated with the lens is changes in color perception. As individuals age, the lens becomes more yellow. This can make it difficult to discriminate between different shades of yellows and greens. Blue items can appear more greenish. Due to this yellowing of the lens, dials, displays, and controls need to be delineated clearly with different shapes, colors, or textures. This form of redundant coding aids older adults in distinguishing between different states of displays without relying solely on color. Subtle differences in shades of greens or blues are not easy for older adults to distinguish and should be avoided on displays and controls.

3.1.7 Useful field of view

One important visual ability that has both peripheral and cognitive aspects is the useful field of view (UFOV). Essentially, the UFOV is the visual area where information can be extracted without moving one's eyes. Since eye movements are slow, the UFOV is critical for rapidly detecting hazards at the side of the road (e.g., a pedestrian who is about to step off the curb).

The UFOV shrinks with age, increasing the need to move one's eyes and head and causing attentional demand and fatigue. This results in older drivers having a form of tunnel vision that can be particularly problematic at complex intersections where attentional demands are high. These complications are compounded if the older driver is fatigued. A decreased UFOV is at least partially responsible for the higher number of intersection crashes experienced by older drivers, although decreased ability to judge distance and speed also play a role.

UFOV is one of the few visual predictors (other than blindness) that has been directly linked to higher accident rates. Fortunately, there is now considerable evidence that the UFOV can be expanded with computer-based training (discussed in Chapter 8).

3.1.8 Macular degeneration and cataracts

The prevalence of macular degeneration and cataracts increases with advanced age. Macular degeneration (damage to the center of the retina needed for sharp focus) is more common among people over age 60, but it can occur in younger adults as well. It can distort central vision, making it more difficult to read road signs and to clearly see traffic and pedestrians.

Cataracts (clouding of the lenses of the eyes) are the most common form of vision impairment for people over age 40 years, and they are more prevalent than macular degeneration and glaucoma. Cataracts can make images appear blurry or out of focus and increase the negative impact of glare at night. About half of older adults do not regularly have eye exams, precluding diagnosis of cataracts and the optimal optical correction important for those who continue driving.

3.1.9 Hearing ability

Nearly two-thirds of adults over age 70 have hearing loss. Gradual hearing loss (or the decreased ability to hear sounds at different frequencies) begins in the twenties and thirties across all frequencies. The rate of hearing loss increases with advancing age and affects higher frequencies before lower frequencies. Sounds need to be increasingly loud to be heard, and particularly high frequencies will be inaudible for older adults. Fortunately the rate of hearing loss is slower for speech frequencies, but even those frequencies are increasingly hard to hear. To be audible to the average 60-year-old, sounds need to be 10–15 dB louder than sounds audible to the average 20-year-old.

The presence of background noise, multiple talkers, and other distractions make listening particularly challenging for older adults. One may not immediately think of good hearing abilities as an essential part of driving or other transportation tasks. But, we use hearing to detect such things as flight announcements in an airport and the approach of oncoming vehicles when crossing a street as a pedestrian. Older adults rely on passengers and other travelers to assist them and find communication in these environments (e.g., noisy airports and city streets) difficult. Further, hearing is used for a wide variety of driving tasks, such as detecting the approach of emergency vehicles and other events outside the car and hearing warnings and alerts inside the car. Older people are known to have poorer "speech-in-noise" perception, which can impair processing of verbally spoken navigational guidance instructions in noisy vehicles.

3.1.9.1 Speech processing

Speech processing difficulties associated with aging are well documented among older adults who often express difficulty in understanding speech, particularly under noisy listening conditions. Traveler waiting areas in bus stations, train stations, and airports are often noisy and contain the presence of multiple sources of irrelevant competing sounds, which exacerbate speech processing difficulties. The extra cognitive effort older adults must expend while processing speech in noisy environments reduces working memory capacity (e.g., making it more difficult to remember strings of navigation commands). That problem can compromise hazard detection while driving. Auditory forward collision warnings in modern vehicles that have been presented well above the hearing threshold of older adults can offset hazard detection difficulties. Processing difficult speech in noise (e.g., noisy background or poor cellular phone connection) utilizes limited working memory resources, which could otherwise be devoted to vehicular control and maintaining awareness of the roadway environment. Therefore, it is essential that auditory route guidance systems be designed to present directions at loudness levels that facilitate processing—currently suggested to be at least 10 dB above ambient background noise. Providing route guidance in a redundant visual format is also important for older people.

The synthetic speech used in many automation systems presents an added challenge. Poor-quality (low intelligibility) synthetic speech requires more cognitive effort to process and further exacerbates the difficulties just mentioned. Fortunately, speech synthesis technologies have greatly improved in recent years, thus reducing intelligibility issues.

3.1.9.2 Sound localization

Sound localization is important in many transportation modes. For example, pedestrians need to be able to detect the direction of approaching cyclists and cars, drivers need to be able to determine the direction of an approaching ambulance, etc. Sound localization abilities begin to degrade in the thirties and become progressively worse with advanced age. Front/back errors in emergency siren localization are a problem for all drivers. Relevant to older people, hearing aids can exacerbate sound localization problems.

3.1.10 Taste and smell

Taste and smell are notably less important to maintaining mobility than vision and audition. However, it is worth noting that older adults will tend to have lower acuity in these senses as well. They can be expected to be less able to detect the presence of hazardous chemicals, such as gasoline or natural gas. Reduced sense of smell makes it more difficult for older adults to detect gas leaks and other potential hazards. Wherever possible, providing a redundant visual alert for these potential hazards will assist the detection capabilities of older adults.

3.1.11 Tactile

Tactile displays are increasingly being used in automobile displays. Virtual rumble strips to signal unintended lane departures from vibrotactors located in the seat pan are just one of the many examples of in-vehicle tactile signaling technologies making their way into the modern automobile. At present, little research has been conducted regarding their feasibility for providing efficient signaling capabilities for older adults. But, in line with existing knowledge of the increased salience needed in visual and auditory modalities for these signals to be detectable to older adults, it is safe to say that tactile signals will also need to be well above threshold sensation levels to effectively signal older drivers. Tactile pulse durations should be at least 50 ms with pulse rates no more than roughly 5–6 pulses per second for stimuli presented to the hand or wrist area. Older adults have considerably less sensitivity on their feet relative to other parts of their body and relative to younger adults; therefore, tactile signals provided to the feet (such as brake or accelerator pulses that have been examined in several recent studies) are likely to be problematic for older drivers.

3.2 Dual-sensory impairment

Dual-sensory impairment (DSI) or having deficits in more than one sensory modality places particular challenges on mobility and overall quality of life. In persons over age 55, DSI is associated with greater fall risk. Either visual or hearing impairments alone can be confused with other disorders and can result in reductions in normal activities and social engagement. However, DSI has a particularly detrimental impact on the ability to carry out activities of daily living (ADLs) and maintain a high quality of life. DSI also increases mortality rates.

People with DSI report mobility difficulties both in the home (e.g., getting in or out of a bed or chair) and outside the home (e.g., visiting friends). Recent research indicates that having impairments in both visual and hearing abilities places older drivers at greater crash risk relative to impairments in only one of the two modalities (Green et al., 2013). However, this finding must be interpreted with caution, because having DSI is also associated with greater rates of comorbid adverse health conditions and greater cognitive decline, particularly for those with low social engagement. It is difficult to disentangle the relative contribution of sensory impairments and other health conditions to social disengagement. Likewise, age-related sensory impairments can be confused with or exacerbate cognitive declines.

Perceptual processing relies on attentional processes, making the relative contribution of each difficult to discern. Suffice it to say here that the degraded or noisier perceptual representation experienced by older adults poses a challenge for higher-order cognitive processes. The more extreme the sensory deficit, the noisier the representation and the more cognitively demanding the information processing task.

3.3 Cognitive changes

Many older adults experience age-related cognitive decline, though tremendous variability is observed across individuals, and cognitive decline is neither universal nor inevitable. However, certain aspects of cognitive function show some levels of decline for most older adults. Age-related declines are routinely found in attention, speed of processing, working memory, and temporal processing and are the focus of discussion here as they relate to mobility.

3.3.1 Attention

There are several varieties of attention, including the ability to focus and to sustain that focus for several minutes or more, as well as the ability to switch focus from one thing to another and to divide focus between two or more things simultaneously. These are all thought to be accomplished by using different forms of attention. Aging impacts all of these processes, which will in turn impact the safety and mobility of older travelers.

Older adults have more difficulties than their younger counterparts searching for and selecting an item to focus on when there are many distractors present. This is true for both visual and auditory stimuli. For example, searching for a particular street name in a crowded city area or listening for a particular directional cue when there are competing voices and noises in the background often proves challenging for older adults. They will be less likely to identify the targeted information, and they will take more time than younger people.

Switching attention from one source of information to another is often found to be impaired in advanced age, as is the ability to divide attention between two or more sources of information. Therefore, processing information from a visual in-vehicle display while maintaining awareness of the surrounding traffic will be more difficult for older adults. Compounding these attentional challenges, older adults have more difficulty inhibiting irrelevant information. For example, if there is a flashing billboard on the side of the road, it will tend to pose more of a distraction to older, relative to younger, drivers. Minimizing cluttered backgrounds in visual displays and decreasing background noise (particularly competing speech) for auditory displays will greatly aid older travelers.

3.3.2 Speed of processing

Aging is accompanied by a decrease in processing speed. This means that it takes longer for older adults to extract information (e.g., read information from a display), and it takes longer for them to switch their attention between displays. This slowing contributes to the "looked without seeing" crashes to which older adults are prone. Slower speed of processing also means older people take longer to access information and to make decisions based on previously learned information. Driving sometimes requires quick responses to avoid sudden obstacles or hazards, and older adults will have more difficulty making these quick decisions. Older drivers are less likely to be able to make split-second decisions to avoid crashes.

Slowed processing speed also means that if a display provides scrolling text, accommodation in the form of slower changes between screens or the ability to self-pace between screens will benefit the older adult. Of particular importance, in terms of safety, displays to be used while driving should be kept terse and low in complexity (discussed in Chapter 5 for signage in work zones). Reduced speed of processing no doubt plays a role in the greater incidence of "looked, without seeing" crashes among older, relative to younger, drivers during left turns at intersections without traffic signals. In addition to difficulties accurately assessing the speed of an oncoming car after one is detected, older adults are more likely than younger counterparts to have insufficient time to visually process the presence of an oncoming vehicle.

3.3.3 Working memory

Working memory is a limited-capacity temporary storage and processing component of memory that appears to contribute to many aspects of complex cognition. Working memory consists of whatever information is actively held in thought at a given moment in time. Older adults have more difficulty than young people in maintaining and manipulating information in working memory. Declining sensory capabilities exacerbate declines in working memory capacity. Reducing working memory load by providing cues, prompts, and visual guidance can aid older adults. For example, a navigation aid that verbally announces an upcoming turn will be strengthened by the addition of a visual reminder of the name of the street and a turn signal cue on the display. A visual depiction of the name of the upcoming street will also serve to clarify an ambiguous verbal cue. Working memory is composed of a central executive component subserved by separate verbal and visuospatial components (Baddeley & Hitch, 1974). While all aspects of working memory tend to decline with advanced age, spatial working memory seems to be particularly negatively impacted by age.

3.3.3.1 Spatial working memory

Spatial working memory is used to hold a goal in mind while performing an unrelated task, such as when we must navigate, find our location on a map, remember where our car is parked, along with a host of other things. Aging appears to take a greater toll on spatial working memory relative to verbal working memory. This commonly observed effect suggests that designs that shift visuospatial demands to verbal working memory may benefit older adults. So, for example, letters or other forms of verbal material in displays may be easier than graphical depictions for older adults to maintain and process in working memory. Most smartphone-based navigation systems provide the option to view directions verbally or graphically, as preferred. Spatial working memory also appears to be an important component of multitasking—a key aspect of many everyday tasks, such as driving.

3.3.3.2 Multitasking

Multitasking, as the term suggests, involves performing more than one task at a time. Drivers must multitask—maintain control of the vehicle's speed, lane position, etc., while concurrently remembering where they are going and scanning the environment for potential hazards. Running errands and navigating through a busy airport also involve multitasking, as one must keep the task goal(s) in mind while executing the individual tasks in an efficient manner. Older adults have more difficulty multitasking than their younger counterparts and therefore are more at risk for inefficient or unsafe performance as individual task complexities increase.

To reduce the demands of multitasking inherent in driving, extraneous tasks and interruptions (e.g., phone calls, advertisements, etc.) should be avoided by older drivers to the extent possible. Older drivers tend to compensate for dual or multitask demands by driving more slowly and maintaining a longer headway. However, multitasking demands still tend to decrease performance on both safety-critical and extraneous tasks.

Preplanning efficient routes and the creation of checklists for running errands and accomplishing multiple sequential tasks can assist older adults and reduce the demands of multitasking during the execution of travel, as well as serve as a memory aid if interruptions result in the forgetting of task goals.

Automating portions of the driving task (e.g., use of cruise control, active lane-keeping, navigation software on smartphones) as well as infrastructure design (e.g., signalized intersections and protected left turn arrows) can reduce both the perceptual and cognitive demands experienced by older drivers. Each of these design recommendations is discussed in further detail in Chapter 5.

3.3.4 Long-term memory

Long-term memory is the term used to describe permanent storage of facts, personal episodes of our personal lives, and skills and abilities (e.g., driving, bicycling) developed across the life span. Long-term memory is less impacted by aging than is working memory. In fact, many studies have shown that crystalized intelligence tends to increase across the life span, even as fluid intelligence, including working memory, decreases. This is good news for older adults who can continue to lead meaningful and rich lives gathering increased information, vocabulary, and memories well into advanced age. Older adults with good health status maintain their ability to remember familiar routes and recognize landmarks as they drive. However, the prevalence of certain diseases, such as Alzheimer's, increases with age, affecting long-term memory and, hence, mobility.

Individuals with Alzheimer's disease may be unaware of the impact of their impairment on their ability to negotiate their environment. It is not uncommon for people in the initial stages of Alzheimer's to continue to drive and then encounter the terrifying experience of getting lost and not having any idea how to get home or perhaps to even remember where they were trying to go. In more advanced stages, a person with Alzheimer's may wander away from home and get lost. There have been several recent accounts of older individuals with a form of dementia, like the Alzheimer's type, getting lost in airports after wandering away from a care provider.

There is little in the way of design changes that can be made to assist those with debilitating forms of dementia. Services have been put in place in many communities where volunteers aid family members in locating lost family members when such events occur. Additionally, some airlines have policies in place for assisting memory-challenged adults. These services may include assisting customers with getting on and off the plane and transferring to a connecting flight.

3.3.5 Decision-making

The slowed cognitive processing rates, reductions in working memory capacity, and declining executive function abilities characteristic of older adults negatively impact decision-making processes. For example, reduced working memory capacity can make comparing choices among multiple options more challenging for older adults. Executive functioning is involved in switching between various tasks and inhibiting irrelevant or distracting information. These age-related cognitive changes are likely part of the reason why older adults are more likely to rely on the use of heuristics, rather than more laborious comparative means of making decisions. For example, when the number of alternative choices increases (complexity increases), older people in particular have more difficulty choosing the optimal choice. Further, older adults are more easily persuaded than their younger counterparts by the number of attributes a given choice provides rather than the importance of individual attributes.

3.4 Physical health

A number of health factors can impact the safety and mobility of older adults. Other than maintaining general awareness of this issue, there is little that can be done in the form of design, so these issues are given only sparse attention here.

3.4.1 Medications

Medication usage tends to increase with advancing age. Many medications impact perceptual and cognitive capabilities, but older adults are often not aware of the potential performance consequences of the medications that they are taking. To combat this knowledge gap, the AAA Foundation launched a website called Roadwise RX, where seniors can log all of their medications and supplements and get customized information regarding how they interact and may affect safe driving performance. Designers should be aware of the impact of medication on performance when conducting usability studies and other forms of testing. There is evidence that commonly used anticholinergic drugs (e.g., Dramamine, Advil PM, Chlor-Trimeton) are associated with cognitive decline and cerebral atrophy in healthy older people (Risacher et al., 2016). Older people should be made aware of that evidence so that they use such drugs cautiously.

3.4.2 Sleep

Driving while fatigued or sleep deprived is problematic at any age. Certain medical conditions like obstructive sleep apnea are more common with advanced age. If left untreated, sleep apnea increases crash risk due to the greater prevalence of daytime sleepiness and reduced alertness. Sleep apnea also is known to increase the risk of Alzheimer's disease, but affordable treatments for sleep apnea do exist.

3.4.3 Illness and disease

A number of other medical conditions that have detrimental impact on safe driving (e.g., dementia, arthritis, diabetes, and Parkinson's disease) also increase in prevalence with advanced age. Medical conditions also can interfere with other forms of safe transportation. For example, people with arthritis may have more difficulty turning their heads during lane changes and in climbing stairs to get on trains and buses. Any condition that negatively impacts health can reduce stamina, making it essential that there are frequent places to rest in public parks, in large airports, and in bus terminals.

The design and implementation of walkable neighborhoods and mixed-use neighborhoods (combining residential and commercial areas) assist older individuals with medical conditions to age in place. Mixed-use areas also tend to increase the physical activity levels of older adults, which can, in turn, slow the progression of many illnesses and facilitate increased health.

3.5 Motor abilities

3.5.1 Range of motion

Flexibility of joints and tendons tends to decrease with age, which impedes both the speed of movement and the range of motion. Older adults have more difficulty turning their heads. This limitation can be a problem while backing up a vehicle because reduced ability to turn the head affects the ability to determine whether a backup maneuver can be accomplished safely. Fortunately, the increased presence of backup cameras (mandated in 2018 and later vehicle model years) can greatly aid older drivers. Similarly, the quick left-right-left head movements necessary to safely negotiate a busy intersection can be impaired by limited head range of motion. Blind-spot indicators available in many newer model vehicles (and available as add-on systems to older vehicles) reduce the need to turn the head to monitor traffic before merging and changing lanes. But, negotiating intersections without the aid of dedicated lane signals (i.e., left turn arrows) can still pose considerable challenge for older people.

Decreased range of motion can also mean that many older adults will have difficulty placing items in overhead bins on buses, trains, and airplanes and retrieving luggage from carousals at airports. These issues are discussed in more depth in a later chapter. But, designs such as reduced height luggage carousals like the one pictured in Figure 3.3 have advantages for older and younger airline travelers alike.

Figure 3.3 Tampa Airport luggage carousel. (Adapted from Jackdude101, -Ownwork, CC BY-SA 4.0, https://commons.wikimedia.org/w/index.php?curid=73753773)

3.5.2 Strength

Aging is generally accompanied by loss of physical strength, due to weakening and atrophy of the muscles. Muscle loss varies widely across individuals but in general begins to occur around age 50 and then increases more rapidly after age 70, and it is more pronounced in women than men. This can make doors harder to open, levers more difficult to pull, etc. Older adults will have a more difficult time carrying weighted objects, and in particular lifting those objects. They will walk more slowly, needing more time to cross intersections. Designing luggage carousels closer to the ground can greatly aid older passengers in airports. Automated door openers, power steering, and other common technological features are also of particular benefit to older adults. These and more design considerations will be presented in subsequent chapters. Decreased muscle strength also impacts balance. Nevertheless, there is considerable evidence that even very old people can increase muscle strength by physical exercise programs.

3.5.3 Balance

Declines in balance are measurable as early as in the fifties. Slips and falls occur more frequently with age and have more severe potential consequences due to increased frailty, slower healing times, and risk of dangerous hip fractures that can be fatal. Trouble maintaining balance is one contributing factor to the increase in fall rates observed among older adults. Many additional factors, including visual impairments, loss of muscle strength, and slower response times, etc., also play a role. Balance and strength training (addressed in Chapter 8) can provide some benefit. Additional design improvements, like ensuring that walkways and stairs are adequately illuminated and handrails are present, can greatly assist older pedestrians. These design recommendations are discussed in Chapter 7.

3.5.4 Psychomotor abilities

Psychomotor abilities refer to the ability to make coordinated movements (particularly arms and legs). Skill in this area is affected by sensory, cognitive, and physical changes. Not surprisingly then, aging is associated with declines in psychomotor abilities. Older adults tend to have more difficulty getting in and out of vehicles, buses, trains, etc., take more time to complete these actions, and tend to be less precise in making coordinated movements. For drivers, this can mean slower and less precise maneuvering around a road hazard. For pedestrians it contributes to reduced walking speeds and the higher incidence of trips and falls. There is little direct research on the impact of reduced psychomotor abilities on mobility, but there is some evidence that overall physical fitness is associated with reduced driving exposure (time spent driving) in older people (Mielenz et al., 2017).

3.5.5 Fitness and training

A number of training paradigms have been developed and examined in recent years with the aim of offsetting age-related declines in sensory, cognitive, and physical abilities. Additionally, training programs aimed specifically at improving the skills of older drivers have been developed. These programs are discussed in Chapter 8. Though clearly not a panacea or cure for aging, many of these programs have met with at least some degree of success. Visual-attention training, hazard perception training, and exercise regimens emphasizing range-of-motion exercises have shown particular benefit in reducing crash risk among older drivers.

3.6 Conclusion

Advanced age is associated with a number of deleterious changes in sensory, cognitive, and motor processes. Many of these age-related changes can be accommodated through the design process. Keep in mind that there is also increased variability in abilities in old age, making strict age guidelines impractical and inadvisable for most purposes. In other words, there is likely to be a larger difference between the performance and capabilities of two 70-year-olds, relative to two 20-year-olds. Do not assume that one design will work for everyone at a certain age. But, the following design recommendations should benefit travelers of all ages, particularly older adults.

3.7 Design recommendations

- Aging is associated with decreased acuity in all sensory systems resulting in the need for signals to be more salient (brighter, louder, more intense) in order to be detectable for older adults. Displays need to be well above acuity threshold to be detectable by older adults.
- Slowing the flash rates to 12–20 per second and/or increasing the contrast and luminance of the flashing source will facilitate detection rates in older adults under most conditions.
- Older adults are more susceptible to glare, which makes it more difficult for them to see displays when competing light sources are present. This can make situations like driving at night particularly problematic. Make sure that rearview mirrors have either a night setting switch that is clearly visible to drivers or use electrochromatic or auto-dimming rearview mirrors.

- Whenever possible, present signals in more than one modality (e.g., both visual and auditory). The use of more than one modality in a redundant manner increases the chance that a person with auditory or visual impairment will detect a critical signal.
- On screens, avoid rapidly scrolling text to accommodate age-related slowing of visual attention or provide the ability to self-pace between screens.
- Present auditory information at least 10–15 dB above speech perception level to reduce the processing demands of information extraction.
- Reduce working memory demands by including verbal rather than visuospatial information where possible and limiting the number of to-be-recalled pieces of information to three or less.
- Ensure that walkways and stairs are well illuminated and include handrails to decrease the incidence of trips and falls.

Recommended readings

Bullough, J. D., Skinner, N. P., & O'Rourke, C. P. 2010. Legibility of urban highway traffic signs using new retroreflective materials (Report). *Transport,* 25(3), 229.

Eby, D. W., Molnar, L. J., & Kartje, P. S. 2009. *Maintaining Safe Mobility in an Aging Society.* Boca Raton, FL: CRC Press/Taylor and Francis Group.

Greenwood, P. M., & Parasuraman, R. 2012. *Nurturing the Older Brain and Mind.* Cambridge, MA: MIT Press.

Schaie, K. W., & Pietrucha, M. 2000. *Mobility and Transportation in the Elderly.* New York, NY: Springer.

chapter four

Older adults on the road

At 3 p.m. every Monday, Maria picks up her granddaughter, Franny, from preschool. Maria has been doing this drive since the start of the school year to help Franny's mom, Anna, who started a new job and is seldom able to get off work in time to pick Franny up from school. Anna drove Maria along the route several times before Maria had to drive the route for the first time. Maria knows the route very well now. Even though she had to get on the freeway to get to the preschool, it was only two exits away, so she did not have to change lanes once she merged in. However, on this Monday, there was a two-car collision that closed the on-ramp to the freeway. Detour signs to the next freeway entrance were set up, but Maria was still uncomfortable with the change in routine. Upon following the detour signs, Maria quickly found herself in a part of town she did not know well. She could not find the next entrance ramp to get back to the freeway and her accustomed route. She drove around a bit, feeling upset. Finally, she pulled to the side of the road and looked in the glove compartment for a map. Although half her mind was on the rapidly passing time, she finally figured out her position on the map and located the entrance ramp a few blocks away. Maria finally safely arrived at the school but was late in picking up her visibly upset granddaughter Franny. Meanwhile, the school had phoned Franny's mother, Anna. Maria belatedly remembered the cell phone that she kept, turned off, in the glove compartment of her car. Franny's mother had tried to reach Maria on that cell phone. Maria got Franny safely home but was rattled by her confusion and delay due to the detour. Later she heard on the evening news that the crash had involved a 79-year-old man and a 37-year-old mother of two. Authorities were still investigating the cause of the crash, but meanwhile, the man had been hospitalized and the female driver of the car and her two children were treated and released with only minor injuries.

4.1 Crash rates and severity

The risks of fatalities in a crash are particularly severe for older people. According to National Highway Traffic Safety Administration (NHTSA) data, both teenagers and older people (aged 55 years and older) are more likely to be involved in fatal crashes than drivers aged 20–54. When people over 70 are involved in a crash, they are more likely to die than are younger people in a crash. When adjusted for miles traveled, fatal crash rates increase after age 70, peaking among drivers 85 and older. This has been attributed to greater susceptibility to injury in older people (e.g., aging is accompanied by increased frailty and slower healing times).

Yet traffic fatalities overall have recently seen a surprising rise across all ages. Highway fatalities declined steadily for five decades, but the first six months of 2016 saw a sharp increase of 10.4% over 2015 (NHTSA, 2016). That increase was in addition to the 7% increase in fatalities seen in 2015 over the 2014 level. These increases are across all ages and represent a troubling reversal of decades of improvement in traffic fatalities. The situation is particularly severe in the United States. The recent "Road Safety Annual Report" found that the United States had a 40% greater vehicle fatality rate than did Canada or Australia. Older drivers particularly are at greater risk insofar as older drivers are keeping their licenses longer and driving more miles than they were even a decade ago. Importantly, effects of age on driving are seen across the adult life span.

The cause of the recent increase in highway fatalities is not fully understood. A portion of the increase in fatalities could be attributed to the improving economy, with people driving more miles. However, the number of deaths as a percentage of miles driven is also increasing (National Safety Council, 2017). About half of fatalities involved occupants who were not wearing seat belts, and a third of fatalities involved drivers who were impaired by alcohol or drugs. However, we need to look for new factors to explain the recent rise in fatalities. Several states have raised speed limits following the repeal of the National Maximum Speed Law. Six states now have speed limits of 80 mph. Texas increased the speed limit to 85 mph in some rural areas. Nationwide about 1500 miles of road now have speed limits of 75 mph or higher. A study conducted by the Insurance Institute for Highway Safety (IIHS), a nonprofit scientific organization funded by insurance companies, concluded that rising speed limits were the cause of 33,000 of the fatalities from 1993 to 2013.

Another factor that could contribute to the rise in fatalities is the increasing "connectedness" of drivers due to the use of cell phones and electronics in cars. However, data on the effect of connectedness on crashes is sparse. A recent study using the SHRP2 (Strategic Highway Research Program 2) database that examined 905 crash events found that the risk of fatality from using a handheld cell phone is 3.6 times higher than baseline driving (Dingus et al., 2016). Other studies have found similar risks of hands-free and handheld cell phones regarding crashing. There is some evidence that older people use cell phones less while driving compared with young people. However, this trend may be changing as baby boomers who are quite familiar with cell phone technology reach old age in increasing numbers. Further, older drivers experience a decrement in driving performance similar to or even greater than that of young drivers when they are conversing on the phone. These decrements are seen in reaction time, following distance, and rear-end collisions.

4.2 Older drivers make specific errors

It is important to note that older drivers are not impaired in all aspects of driving. Rather, older drivers tend to make specific types of errors. Regardless of driver age, intersections account for a disproportionate share of vehicle crashes. Intersections are complex places containing a concentration of information that is critical for the driver (signs, signals, vehicles on conflicting trajectories, etc.). The Federal Highway Administration (FHWA) reported in 2007 that 44.8% of all crashes and 21% of all fatal crashes occurred at intersections. Compared to younger people, older people are more likely to have crashes in stop-sign-controlled intersections, particularly while making left turns. The National Motor Vehicle Crash Causation Survey (NMVCCS) shows that intersections may be particularly challenging for those older drivers who have cognitive, sensory, and/ or motor deficits. The NMVCCS database codes a "critical reason" for each crash. The most common reason for crashes involving older drivers was "inadequate surveillance" (33%). Among older people making that error, 71% looked without apparently seeing (often referred to as "looked without seeing" errors). Importantly, most of these surveillance errors occurred during left-hand turns. Among middle-aged drivers making surveillance errors, only 40% looked without seeing, being more likely not to look at all. Another common error among older drivers was misjudgment of the gap between vehicles. Compared to younger drivers, older drivers were less likely to have crashed due to speeding or overcompensation errors in steering. Also, older drivers were less likely to have rear-end crashes and run-off-road crashes (when vehicle leaves the roadway before crashing).

The problem that older drivers have of looking without seeing may be partially explained by evidence that in intersections gaze patterns of older drivers are different from gaze patterns of young drivers. Older drivers detected road hazards at the same level as young drivers, and showed similar patterns of gaze concentration. However, older drivers were less likely than younger drivers to scan *outside* the planned path of the vehicle before completing a left turn. Automation may provide a partial solution. Starting in 2015, Volvo and Audi offered versions of automated "left-turn assistance," which automatically brakes during a left-turn maneuver if the driver attempts to turn too close to an oncoming car (and during right-hand turns in right-hand drive cars in England and Japan). Such automation could be particularly important in reducing crashes in older drivers who are at greater risk of making errors during left-hand turns.

It is also the case that older drivers who have vision deficits are particularly prone to crashes. In the SHRP2 Naturalistic Driving Study data (behavior and crashes of people driving vehicles equipped with cameras focused on driver behavior inside and on traffic behavior outside the vehicle), the sample of drivers who were over 70 were found to show more near-crash involvement if their "useful field of view" (UFOV) was impaired. Recall from Chapter 3, the UFOV refers to the visual area over which people can rapidly process visual information without moving eyes or head. Actual crashes were also more frequent in those with impaired contrast sensitivity and vision loss in far peripheral fields.

Another solution to the visual problems of older drivers involves better road signage. This is discussed extensively in Chapter 5. In-vehicle automated systems that display road signs on the dashboard could help older drivers who might have missed an important road sign. Such systems display road signs continuously and update them as needed (this is imperfect at present). This technology is also discussed in Chapters 5 and 11.

In summary, older drivers are particularly prone to surveillance errors related to scanning outside the planned path of the vehicle. This is important in crashes during left-hand turns. In addition, older drivers with vision problems are at increased risk of crashing.

4.3 Pedestrian fatalities

We are all pedestrians at one time or another, so it is concerning that pedestrian fatalities have been rising since 2005. In 2015, 15% of traffic fatalities and 3% of traffic injuries were pedestrians. There was a 9.5% increase in pedestrian fatalities from 2014 to 2015 alone. Although children had the highest percentage of pedestrian fatalities (21%), people over 65 had the next highest (19%). Important for regulatory and design purposes, most pedestrians were struck by the front of the vehicle, rather than the rear or side of the vehicle. Most pedestrian deaths are due to the impact of the head on the hood or windshield causing traumatic brain injury (Figure 4.1).

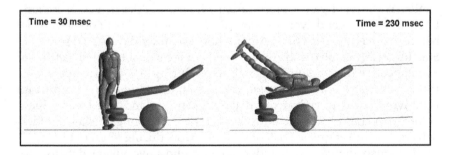

Figure 4.1 Events when a pedestrian is struck by a car.

In Europe, vehicle manufacturers have been required since 2016 to meet a specific standard of pedestrian protection. There is no similar requirement in the United States, although some manufacturers have added pedestrian and bicycle protection to certain models of cars sold in the United States (e.g., Volvo S40 and XC90, Subaru Impreza, and Mercedes). These are pedestrian detection systems that automatically brake the car when a pedestrian or bicyclist is detected by the automation (see also Chapters 6 and 7). Other approaches attempt to reduce the harm to the struck human. Several manufacturers have added external pedestrian air bags on a few models. An approach that meets European standards uses sensors that raise the hood in a crash to lift the pedestrian away from the engine, thereby allowing the person to strike a more forgiving surface. Head injuries are most severe when the energy-absorbing sheet metal of the hood is close to the inflexible cast metal of the engine. A gap of about 10 cm between the hood and the engine allows the pedestrian's head to decelerate and reduces risk of death.

Collisions from backing up into pedestrians are relatively rare, but fatal backup crashes are most likely to involve children. Backup cameras were mandated in passenger cars in the United States beginning in 2018, and that change has the promise to reduce backup injuries and fatalities.

In addition to the harm to a struck pedestrian, there is also harm to the driver involved in terms of emotional cost, insurance cost, and potential loss of independence if a driver's license is suspended. Insurance companies are often unwilling to insure drivers over age 70 who have a record of crashes. There is also potential injury to drivers from hitting large animals (e.g., deer and moose). Several manufacturers offer systems that use infrared sensors for nighttime detection of pedestrians and large animals (e.g., Audi).

4.4 Challenges of navigation

Most drivers who are now over 65 years of age originally learned to navigate using paper maps. However, many older drivers use automated systems providing turn-by-turn navigation, including smartphone-based navigation systems. Although a number of vehicle manufacturers offer navigation systems built into the dashboard for an additional cost, those systems are typically inferior to the navigation systems available without additional cost on smartphones. About 80% of young people use their smartphones for turn-by-turn navigation, but only about one-third of smartphone users older than 65 do so. Older smartphone users who are more educated and affluent are also more likely to use their phones for navigation while driving. This gap will likely change as the technologically sophisticated baby boomers reach age 65 in increasing numbers. The only real disadvantages of smartphone-based navigation systems are the small screen size and the need to mount the phone somewhere on the dashboard.

Although they are less likely to use electronic navigation aids, older drivers might actually benefit more from such aids than younger drivers. It is well documented that the ability to navigate spatially undergoes an age-related decline. Consistent with that, healthy older people are impaired at learning new driving routes compared to young people. Older people have problems forming what are termed *cognitive maps*. There is long-standing evidence that humans and animals form mental maps of their spatial environment that are similar to cartographic maps. Older people have been shown to be relatively impaired at forming and using an *allocentric* mental map of space. Allocentric maps preserve relations between several objects/location in space, regardless of the person's location. However, older people are not impaired at *egocentric* navigation, which uses distance and direction from the location of the person's body.

This evidence that older people are impaired at allocentric navigation suggests that they may obtain a particular benefit from smartphone-based navigation. However, smartphone-based navigation systems are not designed with the older user in mind. Although font size is adjustable on most smartphones, the small screens of smartphones may make larger fonts impractical on maps. In vehicle models from 2016 on, smartphone navigation maps can be displayed on the vehicle's console screen (e.g., Apple CarPlay, Android Auto, etc.), eliminating the difficulty of reading a map on a small smartphone screen while driving (see Chapter 1, Figure 1.3). These maps on the console screen would benefit all drivers, but especially may benefit older drivers who are more likely to have vision impairments.

It is not clear how systems that display navigation maps on the console screen affect safety, as data are not yet available to assess their impact on crashes. However, people are likely to continue to rely on smartphone-based navigation systems, whether the directions and maps are displayed on the driver's phone or vehicle's console. Within a few years, data should become available. However, it should be noted that not all smartphone-based navigation systems work well in rural areas, where older people are overrepresented. Rural areas are not as well mapped by phone systems, and cell reception can be weak. For that reason, older people in rural areas may be slower to use such systems for navigation even though they might benefit when they travel outside their home area.

Nevertheless, the increasing numbers of older drivers on the road argues for optimizing phone-based navigation systems for older drivers. The numbers of people over 65 will double by 2050, largely due to aging baby boomers. Importantly, the baby boomers are more computer literate than earlier aged cohorts. The Bureau of Labor Statistics found recently that 47% of employed people aged 65 years and older used a computer at work. Some 96% of working Americans use new communications technologies as part of their daily life. Therefore, although the numbers of older drivers will be increasing, those drivers will also increasingly be computer literate. Based on that, tailoring navigation software to the sensory problems of older drivers would be prudent as well as potentially profitable. In 2016, there were 111 million people over age 50 in the United States, owning over 80% of the US household net worth (US Census and Federal Reserve data). That suggests that there is a growing market for devices aimed at older people, including older drivers who value their mobility and independence and might be willing to pay for navigation services tailored to their needs.

Basic research will be needed to inform designs required to accommodate and appeal to older drivers. Recent work points to several important factors. Route planning software giving verbal instructions *during* left turns specifically interfered with driving in older but not young people (Paire-Ficout et al., 2016). Most smartphone-based navigation systems allow "heads up" or "north up" displays. As older people are somewhat impaired in allocentric navigation, the "heads up" mode can be an important feature and one that increases the usefulness of electronic navigation systems for older drivers.

4.5 Self-regulation of driving

The older driver population tends to self-regulate their driving for reasons related to their perception of risk and to their recognition of declines in their physical, visual, and cognitive abilities. Older drivers can self-regulate by modifying their travel schedules or by driving less in consideration of weather, traffic, and other environmental issues.

Specifically, older drivers tend to avoid longer trips on interstate highways, reducing their exposure to high-risk traffic conditions including trucks. Older drivers also take into account environmental factors when regulating their driving. There is evidence that drivers over 65 have relatively fewer crashes at night and in cloudy, wet, or snowy weather as they tend to drive less during those adverse conditions. Drivers over 75 reduced driving in poor weather and at night, while fewer younger drivers took those actions. Gender is also an important factor in self-regulation of older drivers, with older female drivers more likely to limit their driving and at a younger age than older male drivers.

Self-regulation of driving by older people can have a negative unintended consequence. The "low mileage bias" refers to the evidence that lower annual miles driven in older people is associated with higher crash rates. This phenomenon appears to contribute to the higher crash rates of older drivers. When odometer data are used instead of self-reported mileage, the low mileage bias is reduced though still present.

This evidence suggests that older drivers should perhaps not avoid driving unless they plan to stop altogether.

4.6 Fatigue

Fatigue and drowsy driving is a known risk factor for crashes. The well-known Virginia Tech naturalistic driving study has monitored natural driver behavior 20 seconds prior to crashes (e.g., Dingus et al., 2016, reporting on the SHRP2 data). That data reveal that drowsy driving behavior increases individual crash risk (odds ratio of 3.4). Self-report confirms those data. A poll conducted by the National Sleep Foundation found that about 60% of US adult drivers say they have driven while drowsy, 37% say they have fallen asleep while driving, and 4% say they had a crash or near crash due to falling asleep while driving. Men between 18 and 29 are most likely to report drowsy driving and falling asleep while driving. Older people are at lower risk of drowsy driving, with 19% of people over 65 reporting drowsy driving. There is no way to confirm these figures, but it is probably safe to assume some degree of underreporting. Data from Australia and European countries who keep statistics on drowsy driving have reported that drowsy driving accounts for 10%–30% of crashes. The lower risk of older people may be due to the self-regulation of driving, particularly at night, as previously discussed.

4.7 Alcohol

Alcohol has a strongly negative influence on safe driving, with intoxicated driving estimated by the NHTSA to account for 37% of fatal crashes annually in the United States (12,000 deaths). What does this mean for healthy older people? As many people now remain employed late in life, the likelihood of work-related alcohol consumption (e.g., "happy hours") is also increasing. Under pressure from Mothers Against Drunk Driving (MADD), a number of states instituted "happy hour" bans in the 1980s to reduce the overconsumption that was encouraged by drink price reductions in the after-work time period when people are likely to also be driving (NHTSA). It is important to note that a driver does not need to be drunk to experience an increased risk of an alcohol-related crash. Even moderate drinking with blood alcohol content (BAC) below the 0.08% standard for prosecution (NHTSA) increases the risk of crashes. Several studies have found that a BAC of 0.04% significantly increases the risk of a crash. Further, the likelihood of a fatal injury from a crash was higher in people whose BAC was elevated but below 0.08% compared to people who had not been drinking. There is evidence that drivers with suboptimal skills are at increased risk of impaired driving under alcohol. Since most older drivers show some skill degradation, this suggests a particular concern when older drivers drink. In a driving simulator study, older adults were more impaired following alcohol consumption than young adults, even with BACs well under the 0.08% limit (Sklar et al., 2014). This suggests that older adults are at greater risk than younger adults when drinking before driving.

4.8 Medication

A majority of people over age 65 take some medication on a regular basis. A physician prescribing a given medication may be ignorant of other drugs a patient is taking, including over-the-counter drugs. This is a potential problem for older people who are likely to be taking a number of medications (polypharmacy). Regarding the effect of medication on driving, certain drugs have been shown to cause drowsiness, but these must be labeled as such by the US Food and Drug Administration (FDA) as they carry a risk for those who take them while operating machines. The FDA also requires prescription labels to indicate when a drug should not be used in conjunction with alcohol. Zolpidem (Ambien) is an example of a drug that strongly interacts with alcohol and is increasingly prescribed to older people for insomnia. Several recent studies have found that opioids (e.g., Tramadol) and sedative-hypnotics (e.g., Ambien) do increase the risk of crashes, especially in people over age 80. The American Geriatrics Society has a list of common practices for the care of older people that both clinicians and patients are urged to challenge when recommended by physicians. Use of zolpidem (Ambien) for insomnia is one of those practices. There is also concern about the use of anticholinergic drugs in common use (e.g., Dramamine, Benadryl, Advil PM, Immodium, Paxil, and Spasmex). It was recently reported that older people taking such anticholinergic drugs on a regular basis have lower memory and attention function and greater brain atrophy compared to older people who are not taking such drugs (Risacher et al., 2016).

4.9 Design recommendations

- Automation with particular benefit for older drivers
  - There exists automation ("left-turn assistance") that applies brakes automatically when a driver turns into the path of an oncoming vehicle. In light of the particular risk posed by such turns for older drivers, such automation could greatly benefit older drivers. However, at present, left-turn assistance systems are offered only in more expensive models of a few manufacturers (viz., Volvo and Audi).

- Automatic emergency braking to avoid rear-end collisions is now widely available in even affordable new vehicles. Use of these systems is highly recommended. However, it is essential that older adults receive information and training on the availability and proper use of these systems.
- Resources for older drivers
  - The American Automobile Association (AAA) has a website aimed at helping older drivers. One feature called "Roadwise Rx" allows people to enter their medications in order to obtain information about how those might affect safe driving in combination. A confidential list can be generated to share with physicians as a starting point of discussion. Another feature is a self-rating tool "Drivers 65 Plus," which poses a series of questions and provides a score and driving assessment based on answers. The AAA also offers "In-Car Brush-up Lessons" for a fee of $80 per hour with certified driving instructors.
  - Several pharmacies have websites that provide information about drug interactions, for example, CVS, Walgreens, and the Mayo Clinic.
  - The AAA also offers well-known pretrip planning ("TripTiks"), both online and paper versions, providing annotated maps and written directions. These show exits, rest areas, scenic routes, construction problems, and other information. While much of this is now available on smartphones, older people are less likely to use smartphone than younger adults.
  - Importantly, there is evidence that certain types of training improve driving performance among older people specifically (Roenker et al., 2003). The benefits include a reduction in at-fault accidents that lasts at least for 2 years in healthy older people. This training is aimed at increasing the speed of processing events in the center of vision and in the periphery of vision. Such training could help older people process information faster. This training is commercially available through the Posit Science Corporation. See Chapter 8 for additional information on design and recommendations for training.
- Unmet needs of older drivers
  - There is a need for trip planning services aimed at the needs of older drivers: information about how to avoid left turns and difficult intersections (e.g., intersections with no traffic lights), nonfreeway routes, and routes with reduced truck traffic. The AAA as well as smartphone navigation tools could address that need in both older as well as very young drivers. Some smartphone navigation systems have a setting that avoids intersections where left turns are difficult.

- There is a need to implement structural solutions shown to reduce crashes during left-hand turns, for example, channelization, J-turns, etc. (discussed in detail in Chapter 5). These would benefit all drivers but would particularly benefit older drivers.
- Alternatives to driving
  - Older people can also reduce their risk of vehicle fatalities by using ride-hailing services like Uber and Lyft. In the US northeast, 41% of smartphone users use their phones to look up public transportation information. Such smartphone use could be helpful for older people in contacting and managing ride-hailing services and in planning bus and rail routes (see also Chapter 9). However, even in the absence of a smartphone, ride-hailing services can be arranged via a website.

Recommended readings

Coughlin, J. E., & D'Ambrosio, L. A. 2012. *Aging America and Transportation: Personal Choices and Public Policy*. New York, NY: Springer.

The FDA on the effect of medication on the ability to drive safely: https://www.fda.gov/downloads/Drugs/ResourcesForYou/UCM163779.pdf

The National Institute on Aging (NIA) has a website containing evidence-based advice for older drivers, including flexibility, cognitive skills, vision, audition, reaction time, and medications: https://www.nia.nih.gov/health/older-drivers#medications

chapter five

Transportation infrastructure

The design of transportation systems can have a major impact on the ease with which older drivers travel safely. In general, driving is a highly visual and complicated task. About 90% of driving information is captured through the eyes so that visual information plays a significant role in driving. Older people typically have poorer visual acuity, contrast sensitivity, and motion detection than young or middle-aged people, with consequences for driving. Crashes are more frequent in older people who have impaired contrast sensitivity and far peripheral field losses. Importantly, good correction of visual acuity may have greater benefits than previously realized. Yet, the majority of older people have outdated prescriptions for corrective lenses. Adjusting for visual acuity eliminates the well-established effect of aging on event-related potential amplitude and latency. Such findings are consistent with the sensory deficit hypothesis of aging that sensory loss impairs cognitive processing due to reduced capacity for rapid information extraction. Considered together, this evidence underlines the need for traffic engineering to accommodate age-related sensory and cognitive losses.

In response to the growing population of adults aged 65 and older, the Federal Highway Administration (FHWA) developed the *Handbook for Designing Roadways for the Aging Population* to provide guidance linking age-related functional decline to a range of traffic engineering treatments. Although the *Manual on Uniform Traffic Control Devices* (MUTCD) governs specifics of all traffic control devices (e.g., lights and road signs), the *Handbook for Designing Roadways for the Aging Population* provides guidance on ways to improve processing of such devices by older people. In this chapter, we do not attempt to replace that valuable handbook. We do attempt to provide insight from new research and to highlight where further guidance is needed. Some of the recommendations in the handbook are based on older research, on small numbers of studies, and often using focus group methodology. Therefore, in this chapter we briefly review standards and focus on innovations as well as problems for which there is less consensus regarding solutions.

5.1 Signs and lights for older drivers

Insofar as the ability to visually detect events declines with age, signage should accommodate that decline. As noted in earlier chapters, aging drivers have decreased contrast sensitivity and therefore need increased levels of light during night driving. According to the Fatality Analysis Reporting System, 49% of all fatal crashes occur at night. Older people also exhibit lower alertness, slower detection in complex situations due to distraction, slower comprehension of messages and symbols on signs, and slower decision-making (*Handbook for Designing Roadways for the Aging Population*).

5.1.1 Signs

Changes that render road signs more conspicuous (both detectable and legible) to older people, especially at night, also render road signs more conspicuous to young and middle-aged people. Therefore, there is no disadvantage in adapting signs for the needs of older people. To that end, considerable research has been aimed at developing standards to promote conspicuity of signs for older people. Conspicuity is a property that emerges from the interaction between sign characteristics (e.g., color and font), sign background (e.g., color and retroreflectivity), scene complexity, and driver expectancy. Expectancy plays a role even for stop sign detection. Following a "Stop Ahead" sign, stop signs are detected faster in all drivers due to heightened expectancy.

Retroreflectivity is a measure of how efficiently material returns incident light at a given distance, angle, etc. between the sign, the vehicle, and the driver. Based on research sponsored by the FHWA, Mace and colleagues developed recommendations on the retroreflectivity needed to make yield and stop signs conspicuous for both young and older drivers. These recommendations were based on a model that predicts retroreflectivity levels needed for visibility as a function of (a) driver age and visual acuity, (b) the need for the driver to take action in response to the information on a given sign, (c) a given driving situation (city or highway), (d) given operating speeds ranging from 10 to 65 mph, and (e) signs of varying size and placement (shoulder, overhead). These recommendations are detailed in the *Handbook for Designing Roadways for the Aging Population*. The MUTCD was most recently updated in 2009 to change the standards for minimum retroreflectivity.

Luminance is another issue that is particularly important at night when fatal crashes are more likely. Luminance is the amount of light reflected from an object, in contrast to retroreflectivity, which is a property of the object's material. Greater retroreflectivity is generally associated with greater luminance, but that can depend on the source of illumination and the angle from which the object is viewed. The poorer visual acuity typical of older drivers means that they require greater luminance to detect and read signs. The acuity of older people degrades more under low levels of luminance than acuity of younger people. Several recent studies have found that the ability of older people to recognize objects while driving is best predicted by a combination of mesopic (low light) acuity with photopic (bright light) acuity (Wood & Owens, 2005). This has implications for how vision is tested for driver licensing, which is currently done only under photopic (bright light) conditions.

Work zone crashes have increased in recent years, and the safety of workers in addition to the safety of vehicle occupants is at risk. Work zones present a particular problem for signage. Conspicuity needs are different for work zone signs insofar as the signs must be readable (legible), in contrast with stop and yield signs that are recognizable by shape and color alone. The information provided by work zone signs is often specific and novel. Legibility of words on work zone signs at night is a particular concern for older drivers. Several studies comparing young and older drivers have found the older drivers were able to read words on signs much later than the young and middle-aged drivers. Based on this, the *Handbook for Designing Roadways for the Aging Population* recommends that redundant signs be used in work zones.

Retroreflectivity of work zone signs has been investigated in trials by Chrysler and colleagues. They compared the type of retroreflective sheeting (material used to make the sign) as well as the font, and font color in both young and older drivers. Although the specific fonts did not have an effect on legibility, microprismatic sheeting (the reflective layer of which consists of microscopic reflectors in cube-corners) was more legible at night than the standard retroreflective sheeting. Regarding daytime legibility, fluorescent orange sheeting results in greater sign legibility than standard orange sheeting. The benefit was greatest for older drivers.

Changeable (variable) message signs (Figure 5.1) are commonly used on interstates and in work zones. Standards for their use are addressed in detail in the *Handbook for Designing Roadways for the Aging Population*. As older drivers exhibit slower comprehension of messages and symbols on signs as well as slower decision-making, it is important to consider their needs. Empirical studies have found the following regarding changeable message signs:

- Increasing luminance benefits all drivers regardless of age.
- When the message signs are on the shoulder, elevating them over the roadway reduces visual blockage of the signs by traffic.
- Lighter colored letters on a dark background (positive contrast) were easier to see than the reverse.
- Red-on-black signs were not read as easily by older drivers as yellow-on-black or white-on-black signs.
- Redundant signs promoted compliance.
- The current standard reading time of signs was acceptable for older drivers.

Figure 5.1 Changeable message sign. (This file is licensed under the Creative Commons Attribution 3.0 Unported license.)

- Short messages (fewer words) reduce working memory load on older drivers.
- Messages that involve more than one phase (message split in time) should appear in no more than two phases.
- No effects of age were found on reading time or on comprehension of static compared to flashing messages.

5.1.2 Driver behavior at traffic signs

About 40% of fatal crashes occur at stop-sign controlled intersections (reported by the Insurance Institute for Highway Safety [IIHS]). In most of these crashes, the driver initially comes to a stop but then fails to yield. Older drivers disproportionately fail to stop at stop signs and are disproportionately involved in crashes at stop-sign controlled intersections. This may be related to the "looked without seeing" tendency observed in older people during left turns (see Chapter 4). The well-studied age-related reduction in the "useful field of view" (UFOV, the visual area over which information can be rapidly extracted in one brief glance as discussed in Chapter 3), impairs detection of road signs. An in-vehicle monitoring study found that older rural drivers run stop signs more often than older urban drivers (Keay et al., 2009). However, young drivers also run stop signs more than middle-aged drivers, and their behavior is not likely due to vision problems. To increase compliance, some states have added flashing lights (embedded LEDs) to stop signs. That addition has been found to increase full stops and reduce the number of vehicles that fail even to slow for stop signs. When a stop sign is not visible at the standard distance set by the American Association of State Highway and Transportation Officials (AASHTO), then the addition of "Stop Ahead" signs is recommended. Also recommended are supplemental warning signs below the stop sign at two-way stop-controlled intersections ("Cross Traffic Does Not Stop"). If people are deliberately failing to stop at stop signs in rural areas where traffic is low and lines of sight are good (as some data suggests), then perhaps a public education campaign and/or increased enforcement would be indicated. In Section 5.2, we discuss structural treatments to address the serious problem of crashes at stop-sign controlled intersections.

5.1.3 Traffic lights

The *Handbook for Designing Roadways for the Aging Population* makes a number of detailed recommendations for traffic lights. Rather than repeating that information, we highlight two of those with particular relevance for older people.

A new recommendation for traffic lights involves the addition of a 1–3 inch retroreflective border around the backplate of traffic lights to enhance the visibility of the signal to reduce red-light running. This border is now listed as an option in the 2009 MUTCD. The FHWA named retroreflective backplates one of their "Proven Safety Countermeasures."

Another recent change in signage recommendations concerns "permissive" left turns, left turns that are permitted only after the driver yields to oncoming traffic. As of 2006, the flashing yellow arrow was approved for use for permissive left turns and added to the 2009 MUTCD. This recommendation change was made based on evidence summarized in the *Handbook for Designing Roadways for the Aging Population* that (a) left-turn crashes are relatively common and often severe, particularly for older drivers, and (b) older people were often confused when a circular green light indicates a permissive left turn. The recommended arrangement of traffic lights has also recently changed. The handbook now recommends a horizontal arrangement of lights, with one signal head per lane (Figure 5.2).

5.1.4 Automatic road sign detection

A number of car manufacturers have developed automation that detects traffic road signs and displays that information on the dashboard (Figure 5.3). The early systems recognized speed limits only, but more recent road sign detection systems can also detect other signs such as overtaking restrictions.

Figure 5.2 Horizontal arrangement of signal lights, with one light per lane. (This file is licensed under the Creative Commons Attribution 2.0 Generic license.)

Figure 5.3 A road sign automated system in a vehicle showing the current speed limit. (This file is licensed under the Creative Commons Attribution-Share Alike 4.0 International license.)

Such systems are becoming more common, being recently offered by a wide range of car manufacturers (e.g., Volkswagen, Nissan, Volvo, and Honda).

5.2 Intersection assistance for older drivers

Intersections are a major concern for older drivers who are overrepresented in crashes occurring in intersections. According to the IIHS, about 40% of fatal crashes of people aged 70 and older occurred at intersections. Not surprisingly, older drivers are more vulnerable to injury from crash forces. The FHWA's *Handbook for Designing Roadways for the Aging Population* provides guidance on the design of intersections to accommodate the particular needs of older drivers with the goal of reducing intersection crashes. The handbook describes both "proven practices" and "promising practices" that involve a range of road treatments from widening lanes to conversion of intersections to roundabouts.

Left turns in intersections are a particular problem for older people, as we discussed in Chapter 4. Roadway design can reduce the problems that all drivers, but especially older drivers, have with left turns. Structural solutions that avoid direct left turns are displaced left turns and median U-turns (see *Handbook for Designing Roadways for the Aging Population*). However, a better long-term solution is to eliminate left turns altogether by using roundabouts, J-turns (Figure 5.4), and jug-handle turns (Figure 5.5). These are all FWHA "proven safety countermeasures," described in some detail as follows.

Roundabouts are a good alternative to stop-controlled intersections. Roundabouts eliminate left turns, increase traffic flow, and have been found to be the safest type of at-grade intersection. However, as roundabouts are

Figure 5.4 A two-way stop controlled intersection converted to a J-intersection. At the stop sign on the minor route, only right turns are allowed. Achieving a left turn from the minor route requires a right turn at the stop sign onto the major route and then a U-turn at the median. To achieve crossing of the major route, after making the U-turn, drivers make a right turn back onto the minor route. Underpasses are shown, but J-intersections are often implemented with level roads (e.g., http://contribute.modot.mo.gov/central/major_projects/JTurns.htm). (Retrieved from https://commons.wikimedia.org/w/index.php?curid=980455)

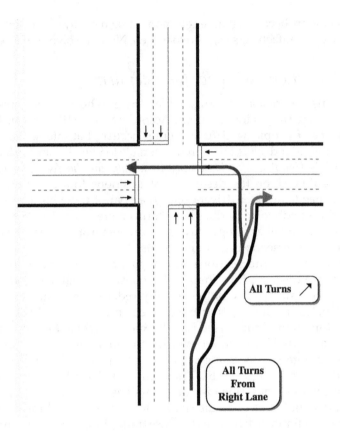

All Turns ↗

**All Turns
From
Right Lane**

Figure 5.5 A jug-handle turn that avoids direct left-hand turns from the major road onto a minor route (crossroad). Drivers exit the major route by a ramp on the right side of the road. Drivers then turn left from the ramp onto the minor route before the intersection with the major route. (From Evan Mason – Own work, CC BY-SA 4.0, https://commons.wikimedia.org/w/index.php? curid=50659854.)

relatively new in the United States, many drivers have limited experience with them. This suggests that signage in roundabouts is important, especially for older people. The *Handbook for Designing Roadways for the Aging Population* makes recommendations on signage in roundabouts. However, these recommendations are based on only one study that used focus groups and therefore should be viewed skeptically.

Overall, highway deaths are higher in rural compared to urban areas. In 2015, the IIHS reported that more than 50% of crash deaths occur in rural areas, although only 19% of the population lives there and 30% of vehicle miles are driven there. One contributor to these crash deaths is the trend of converting high-speed multilane divided rural roads into what are termed *rural expressways*. These rural expressways typically have at-grade intersections with stop signs rather than ramps in order to allow access from crossroads. In-vehicle monitoring at stop-sign controlled intersections of such rural expressways revealed that older rural drivers run stop signs more often than older urban drivers. Failure to yield and stop at a rural expressway would be particularly risky for drivers making left turns and for drivers crossing the expressway.

How could infrastructure address this problem of rural expressways? The least expensive solution would involve additional signage such as "stop ahead" and "cross traffic does not stop" signs. However, to the extent that the failure to stop is voluntary rather than perceptual, other solutions would be needed. To address the problem of crashes at left turns at intersections, the FHWA promotes "Proven Safety Countermeasures" such as J-turns (Figure 5.4) and jug-handle turns (Figure 5.5). States typically respond to problematic intersections that have high crash rates by implementing several changes at once, precluding systematic investigation. A National Academy of Sciences study described 10 rural expressway intersections where different solutions were applied, including J-turns and jug-handle turns (the FHWA terms J-turn intersections as "U-turn-based intersections" (FHWA, 2014; and see Recommended readings). J-turns and jug-handle turns have the advantage of eliminating (a) left turns onto major routes and (b) direct crossing of major routes from minor routes, both associated with high crash rates. Data obtained before and after conversion of seven of these intersections were analyzed, showing a statistical benefit of the conversion. However, there were a number of factors that could not be controlled. A separate study of conversion of five different intersections from two-way stop controlled to J-turn controlled in rural expressways in Missouri resulted in a 53.7% reduction in injury and fatal crashes after conversion.

5.3   Problem areas of older drivers

5.3.1   Gap detection and surveillance

Age appears to reduce the ability to judge the gap between vehicles approaching intersections. The National Motor Vehicle Crash Causation Survey (NMVCCS) database of serious crashes found that gap misjudgment errors accounted for 6% of crashes among older people. As discussed in Chapter 3, older drivers' visual acuity even after correction is typically suboptimal, so they are less able than younger drivers to judge imminent collisions. There is evidence that during left turns when making decisions about whether a gap is safe, older drivers paid more attention to judged distance than to judged speed.

Age also appears to affect surveillance during driving. Maintaining safe driving requires persistent and accurate scanning of the environment to obtain critical information. Older drivers make fewer glances toward their peripheral visual field than toward central vision, which can result in more search errors. The NMVCCS database of serious crashes found that the most common reason for crashes involving older drivers was "inadequate surveillance" (33%). Among older people making surveillance errors, most (71%) looked without seeing. Among middle-aged drivers making surveillance errors, only 40% looked without seeing, being more likely to not look at all. Further, as described in the following text, older drivers were less likely to scan outside the planned path of the vehicle before completing a left turn.

Age-related differences in visual scanning have been observed as drivers negotiate intersections. Compared to younger drivers, older drivers:

- Do not use the full scanning range
- Tend to check fewer separate areas
- Check the rearview mirror less often

5.3.2 Run-off-road crashes

One limitation of the *Handbook for Designing Roadways for the Aging Population* is that it does not address the serious problem of vehicles leaving the roadway. Over half of all fatal crashes occur in what is termed "run-off-road" crashes. Any factor that affects risk of crashes is especially relevant to older people insofar as they are more likely to die in traffic crashes. When a driver runs off the road to the left on an expressway with a traversable median, there is the risk of encountering opposing traffic. Because of that, such cross-median crashes are among the most severe crash types.

One structural solution to run-off-road crashes is a "safety edge," which eliminates the sharp drop-offs at the edge of lanes that are caused by modern road construction methods. Vertical pavement edges may have contributed to 18%–25% of run-off-road crashes on rural paved roads with unpaved shoulders during 2002–2004. The FHWA has a "Safety Edge" initiative to promote formation of a 30° angle in the edge during road paving projects.

Another structural solution involves placing barriers in the median to prevent vehicles crossing the median to encounter opposing traffic. The federal government does not require median barriers, and standards for placing median barriers were developed in the 1970s and have never been revised. Yet, there is evidence that a large portion of cross-median crashes occur outside the regions recommended for barrier placement by the existing standards. Even expert panels do not agree on standards for median barrier placement. As a consequence of the lack of standards, the rules for placing median barriers vary by state. For example, in Florida the state requires a median barrier if the median is less than 64 feet wide. In California, median barriers are recommended for medians less than 75 feet wide, but concrete median barriers are specified when medians are less than 20 feet wide.

Cost is a concern to states regarding the placing of barriers. Most states consider crash history in determining whether the cost of a median barrier is warranted at a given location. Recently, cable barriers (steel cables connected by posts, see Figure 5.6) have come into wider use due to their lower cost than concrete barriers or guardrails. A number of states have reported sharp reductions in cross-median fatalities on highway segments where they installed cable barriers in expressway medians that previously had no barriers. However, cable barriers were not designed to stop heavy tractor trailers, so they can only be a partial solution and may not be optimal on roads with heavy truck traffic. Due to the severity of cross-median crashes and the greater fatality rates for older people involved in any type of crash, development of new standards for median barriers is important.

Another solution to cross-median crashes would prevent vehicles leaving the road in the first place. To that end, car manufacturers are beginning to address this "run-off-road" problem. Several Volvo models

Figure 5.6 Cable barriers are increasingly used in place of guardrails to slow vehicles leaving the roadway of expressways. (By Joel Torsson (Leojth) – Own work, Public Domain, https://commons.wikimedia.org/w/index.php?curid=2096153)

now offer "Run-off mitigation" automated systems that will actively steer the car back on the road and also apply the brakes, if needed.

5.3.3 Lane change errors

Another major concern is the problem of an older driver merging into traffic on fast-moving roads (such as expressways), and of changing lanes while also focusing attention on other vehicles and objects. Changing lanes is among the most dangerous driving behaviors overall, with over 250,000 crashes attributed to lane changes each year in the United States. Examination of national databases revealed that older drivers are disproportionately involved in crashes attributable to highway lane changes. Age-related changes in the ability to control attention have been found to be a stronger factor in lane change errors than age-related changes in vision or stress management (Munro et al., 2010). Consistent with that finding, training aimed at control of attention in older drivers does appear to reduce risky maneuvers (discussed in Chapter 8).

Although there is no solution for lane change errors in terms of road treatments, car manufacturers now offer automated systems that have the potential to reduce lane-change crashes (described in Chapter 6). Automated blind-spot monitoring systems typically provide a warning light near the side-view mirrors when a vehicle traveling in an adjacent lane is approaching the driver's blind spot and is currently in the driver's blind spot. These systems are available as optional equipment in new vehicles (for an additional charge) from a range of manufacturers (e.g., Volvo, Honda, Toyota, etc.). Blind-spot detection systems can also be purchased as aftermarket systems, and some of these get high ratings from *Consumer Reports*. An inexpensive solution involves affixing a convex mirror to the side mirrors. These mirrors, similar to the convex mirrors on tractor trailers, allow the driver to see the area of the blind spot. These mirrors can be purchased online from a range of sources.

There are also automated systems that help keep the driver from accidently drifting out of the lane. Newer cars commonly now offer automation with an auditory alert to warn the driver when the vehicle starts to leave the lane—"lane departure warning." Newer "lane-keeping assist" systems actively steer to keep the vehicle in the lane. All of these can be helpful to older drivers who may have attention problems, as previously mentioned. Older drivers also often have reduced neck mobility so that turning their head to scan for cars in adjacent lanes is slower and more difficult.

Automated systems that could help reduce dangerous lane change crashes are "rear collision warning" systems offered by a few manufacturers. Such systems activate the flashers to alert both the driver and a vehicle approaching very fast from behind. The system also tightens the seat belts and activates a whiplash system. If the car is stationary, the brake is activated to reduce forward momentum during the crash.

5.4 Older drivers and autonomous systems

For several of the problem areas of older drivers as previously discussed, we referred to automated systems designed to reduce specific types of crashes. An important question arises about whether older drivers would have difficulty using such systems. Automatic emergency braking is now widely available in even very affordable new passenger vehicle models and has been found to sharply reduce rear-end accidents, as described. The driver does not need to be aware of the presence of automatic emergency braking systems for them to be effective. Such systems do not stop the driver from braking, but apply brakes only if the driver does not. Similarly, drivers in vehicles with rear-autobrake systems (monitors vehicles approaching from behind) experienced lower numbers of back-up crashes. Therefore, automatic emergency braking systems would be very useful for older adults. Particularly relevant to older people is the automated "left-turn assistance" described in Chapter 4. That automation monitors oncoming traffic during a left turn and brakes automatically if the driver attempts to turn when the gap is judged by the automation to be insufficient. In light of the evidence that older drivers were more likely to have crashes at stop signs, particularly during left turns, such automation has the potential to reduce such crashes in older drivers. Parking assist systems would also be useful for older drivers who may have limited neck mobility and therefore trouble turning to see behind their vehicle during parking maneuvers. Some of these systems use visual and auditory alerts to guide the driver while the driver maneuvers the car. Other systems take over some or all steering and acceleration functions during parking.

Other automated systems might be more difficult for older drivers to use. "Lane-keeping assist" automation actively steers the vehicle back toward the center of the lane. In some models this results in a distracting "ping-pong" effect as the car moves back and forth between the lane lines. The newer "lane centering" technologies steer more smoothly to keep the vehicle in the center of the lane. These systems also depend partly on the clarity of the road lane markings. The intermittent nature of current lane-keeping systems might make them confusing to older drivers. Another type of system that might be confusing to older drivers is that designed to allow limited self-driving, though without lane changes (e.g., "Volvo Pilot Assist," "Cadillac Super Cruise"). These systems brake and steer automatically but stop working if the driver is judged by the automation not to be sufficiently engaged in driving (hands not on the wheel or eyes not on the road). Some of these systems will coast to a stop and activate flashers if the driver is judged not to be alert and responsive. These systems could be confusing because they depend on road type, ambient traffic, and lane line clarity. The intermittent availability (indicated by changing icons on the dashboard) could be distracting in some systems as the driver must notice the disengagement and then reengage the system after it turns itself off. A few high-profile crashes have occurred when the lane-centering systems were confused when a road divided or lane lines were not clear *and* the driver was not monitoring the steering.

It should be noted that all of these automated driver safety systems can be turned off on the dashboard. That could be an advantage for older drivers who might find certain systems annoying or confusing. Automatic emergency braking can also be turned off, but that requires going through several menus in most vehicles equipped with such systems.

5.5 Heat mitigation at bus and rail stops

Older people who use bus and rail for transportation are at greater risk than younger people from the high ambient summer temperatures that are common in a number of US cities. Most bus riders are employed adults between the ages of 25 and 54, with 7% being over 65 years of age. According to the Centers for Disease Control and Prevention (CDC), older people are more susceptible to heat stress than younger people as they are more likely to have chronic medical conditions and are more likely to be taking medications that affect normal physiological responses to high heat. This can be a particular problem in cities. The higher ambient temperatures in urban compared to rural areas have been termed the *urban heat island* effect. The deaths of a large number of older people in Chicago during the 1995 heat wave spurred research into ways to mitigate urban heat islands.

Several cities in the southern United States have developed innovative ways to improve the comfort of passengers waiting at bus and rail stops by reducing ambient heat. Based on empirical research conducted at Arizona State University, the city of Phoenix, Arizona (average summer temperature over 100°F), uses shade structures, vines, and trees to cool outdoor commuter rail stops. These measures benefit all riders, but especially older riders, by mitigating heat at outdoor bus and rail stops. Those innovations could be used more broadly in other cities with high summer temperatures.

5.6 Highway-rail grade crossings

Highway-rail grade crossings (where roads and rail lines intersect) are the scene of a number of rail-vehicle fatalities every year involving pedestrians and vehicle occupants. However, older people are most vulnerable. The most recent data from the Federal Railway Administration indicates that in 2016,

994 pedestrian casualties occurred at rail crossings. Regarding rail-vehicle collisions, of the 25,945 highway-rail crossing accidents in the United States between 2002 and 2011, data normalized by miles traveled showed that the age groups at risk were those under 19 and those over 70 years of age. Those over 70 were at greatest risk. Further, the proportion of older drivers in highway-rail level crossing vehicle-train crashes at night is greater than the proportion of vehicle-only crashes at night. This suggests that reduced visibility of signals at night is a factor in crashes at level crossings.

Most highway-rail grade crossings have passive controls, and most fatal crashes at railroad level crossings occur at passive crossings. The *Handbook for Designing Roadways for the Aging Population* addresses the problem of crashes at highway-rail grade crossing in great detail, noting long-standing research showing that people commonly do not yield at the standard railroad crossing "crossbuck" sign. Further, the recommended height of the sign has been found to be too high to be optimal for headlight beam patterns. Based on work comparing young, middle-aged, and older drivers, recent changes to the MUTCD in 2009 involve the addition of retroreflective material to the crossbuck and the addition of either yield or stop signs to the crossbuck.

5.7 Ambient lighting

Lighting at night is important for driving safety. Road crashes are more frequent and severe at night compared to daytime crashes. Young drivers have disproportionately more crashes at night than older drivers, perhaps due in part to decisions by older drivers to avoid driving at night. The FHWA's handbook reviews the extensive evidence that illuminating intersections reduces crashes. Due to the known reduction of contrast sensitivity with aging, greater brightness of signs and stripes is recommended to accommodate aging drivers. Drivers over the age of 70 are more likely to be involved in fatal crashes from wrong-way movements. Wrong-way movements are associated with intersections with low lighting and restricted sight distance, especially where exit ramps intersect surface roads. Older drivers also show greater susceptibility to glare, which can be a particular problem in work zones where drivers are exposed to headlights of opposing vehicles. It should be noted that the guidelines for traffic signal characteristics of color, size, and function are largely based on a single study published in 1966.

5.8 Design recommendations

The FHWA has recently recommended use of what are termed *proven safety countermeasures* (https://safety.fhwa.dot.gov/provencountermeasures/). These are relatively low-cost means of reducing crashes and injuries:

- Roundabouts to avoid left turns
- Conversion of stop-controlled intersections to J-intersections or jug handle intersections to avoid left turns
- Safety edges on roads to eliminate a sharp drop-off at pavement edges
- Use of retroreflective backplates on traffic lights
- Pedestrian crossing islands and refuge islands (see Chapter 7)
- Road diet (see Chapter 7, which converts four-lane, undivided roads to three-lane roads including a two-way left-turn lane)

In addition to recommendations detailed in the *Handbook for Designing Roadways for the Aging Population*, we make the following recommendations:

- Urge optimal refractive correction for older drivers and pedestrians (e.g., wavefront technology can determine acuity errors by measuring the way that light waves travel through the eye). Poor visual acuity impairs perception and cognition and is a risk factor for crashes.
- Use more extensive vision testing before licensure of older drivers to include testing for mesopic (low light) vision in addition to the standard testing of photopic vision.
- Add flashing lights to stop-sign intersections, including on rural expressways where running stop signs is a problem.
- Apply convex mirrors to side mirrors to reduce the blind spot.
- Increase the use of median barriers on expressways to prevent cross-median crashes.
- Provide heat mitigation interventions near outdoor bus and rail stops in cities with high summer temperatures.
- As older drivers are more susceptible to crashes, develop a public interest campaign aimed at informing older people about benefits of safety automation systems now available in many affordable vehicles. Automated vehicle systems are also discussed in Chapters 6 and 11.
- Develop and provide training aimed at improving driving skills in older drivers (see Chapter 8).

Recommended readings

Federal Highway Administration (FHWA). 2014. *Median U-Turn Informational Guide* (FHWA-SA-14-069). Washington, DC: US Department of Transportation. Retrieved from https://safety.fhwa.dot.gov/intersection/alter_design/pdf/fhwasa14069_mut_infoguide.pdf; and see specifically: https://safety.fhwa.dot.gov/intersection/innovative/uturn/

Federal Highway Administration. *The Handbook for Designing Roadways for the Aging Population.* Retrieved from https://safety.fhwa.dot.gov/older_users/handbook/

Federal Highway Administration. *The Manual on Uniform Traffic Control Devices (MUTCD).* Retrieved from https://mutcd.fhwa.dot.gov/kno_2009r1r2.htm

chapter six

Driver vehicle interfaces and older adults

Advanced features in the modern automobile have the potential to greatly enhance vehicle occupant safety. Many systems are designed to enhance driver safety and are referred to as advanced driver assistance systems (ADASs). Other systems, often referred to as infotainment systems, are primarily designed for comfort, entertainment, and convenience, but they may also help reduce stress and mental workload. All of these advanced systems may be of particular benefit to older adults, if designed in accordance with older driver perceptual and cognitive capabilities and if drivers understand the system's capabilities and limitations. In this chapter we examine the design of the driver vehicle interface (DVI). The DVI includes the primary instrument cluster, such as the speedometer, fuel gauge icons, automated system icons, as well as all other controls and displays that the driver interacts with when operating the vehicle. The advent of ADASs in vehicles certainly increases the cognitive load on the driver, making it important that the DVI communicate well with older people. Our examination of the DVI will include (a) designing to enhance detection, comprehension, and implementation of display information; (b) the impact of using different modalities of information presentation; and (c) other approaches aimed at reducing visual, cognitive, and physical load, particularly for older adults. We also discuss recommendations or design suggestions for specific vehicle technologies, focusing particularly on those designed to enhance safety.

6.1 Display modality

6.1.1 Visual

Visual displays are the most common display modality currently used in vehicles. The Federal Motor Vehicle Safety Standards (FMVSS) require that all vehicles made after 2011 have at least 32 vehicle states indicated to the driver via visual displays. Many of these displays must be illuminated so that they are clearly visible during night driving. Examples of key displays that must be visible at all times include the speedometer, fuel level, engine coolant temperature, and engine oil pressure. The FMVSS standard No. 571.101 for controls and displays also mandates that specific symbols be used for many of the displays. These guidelines can be found online by searching the US Government Publishing Office website (https://www.gpo.gov/fdsys/).

When they are well designed, visual displays can quickly and efficiently impart considerable amounts of information to the driver. However, since the task of driving places heavy visual demands on the driver (e.g., to maintain lane position and monitor traffic and other potential road hazards), best practices in visual display design facilitate the transfer of information between display and driver. We discuss important human factors design guidelines for visual displays that will benefit all drivers, but particularly older drivers. See the list of recommended readings at the end of this chapter for additional guidelines and further discussion on each of these guidelines.

6.1.1.1 Clutter
Clutter in displays is a particular problem for older drivers. Clutter is experienced when there are too many objects in a display and when objects are too close together, disorganized, or too similar in appearance. Clutter should be avoided in displays as it impairs visual search and selection accuracy. To minimize clutter, avoid placing nonessential visual displays in the prime visual search space. Prime visual search space in a vehicle can be defined as within the driver's useful field of view and major forward field of view locations (e.g., the front windshield providing information on the roadway directly in front of the forward-moving vehicle, the primary instrument cluster including the speedometer and the rearview mirror), each of which can be accessed easily with a single glance.

6.1.1.2 Perceptual grouping
Displays benefit from attention to perceptual organization and to grouping principles as these can facilitate understanding and ease of selection. Color coding can be used to indicate controls or displays that belong to a single system. Gestalt principles of similarity and proximity are additional cues for perceptual organization that can aid display design. For example, in Figure 6.1 items that are similar in appearance will tend to be grouped together perceptually. Proximity is another Gestalt principle that aids perceptual organization. In Figure 6.2, the columns of circles that are closer together will be perceived as belonging to the same group.

To see an illustration of the use of these two principles for an in-vehicle system, Figure 6.3 shows a grouping of visual displays for controlling the infotainment system and climate controls.

6.1.1.3 Proximity compatibility principle
The proximity compatibility principle (PCP) indicates that if information from two displays must be integrated (e.g., posted speed limit display and actual speedometer), then the displays should be located in close proximity. Conversely, if they present unrelated information, the

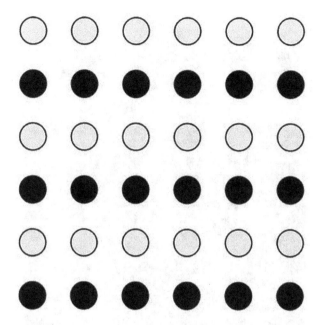

Figure 6.1 Gestalt principle of *similarity* aids perceptual organization. The circles tend to be perceived as rows due to the similarity of open and colored columns circles. Similarity can be used to indicate which displays and controls are associated with a single system.

Figure 6.2 Gestalt principle of *proximity* leads most people to perceive these circles as three columns of circles on the left and two columns on the right. (Adapted from https://commons.wikimedia.org/w/index.php?curid=3961100)

displays should be clearly separated to avoid confusion. Adherence to the PCP is particularly important for older adults, since as discussed in Chapter 3, older adults take longer to extract information from visual displays and longer to switch their visual attention back and forth between displays.

Figure 6.3 Gestalt principles of both *similarity* and *proximity* are used as cues for perceptual organization to distinguish between the controls associated with the infotainment center and those of the climate control system. (Personal image of Carryl Baldwin.)

New models of vehicles with automated systems require space for icons on the dashboard/instrument panel showing the state of those systems. The advent of automated systems requires some rearrangement of the dashboard with the speedometer moved from the center of the display in some models to allow adaptive cruise control and lane-keeping icons to appear in the center. It is important that these dynamically changing icons be as salient and discriminable as possible. In contrast, blind-spot detection system signals appear in the periphery, near or on the side mirrors.

6.1.2 Auditory

Auditory displays have a number of benefits over visual displays that can greatly benefit all drivers, but particularly older drivers. Auditory signals are omnidirectional, meaning that they can be perceived regardless of where a person is looking. Providing information in the auditory modality allows a driver to keep his or her eyes on the road, thereby reducing visual switch time costs. For example, providing verbal navigational guidance messages allows the driver to keep his or her eyes on the road and scan the environment for both hazards and street signs. Additionally, because auditory information can be heard regardless of where a person is looking, the auditory modality is particularly well suited for providing time-critical information such as collision avoidance alerts. However, auditory displays (e.g., verbal directions and collision warnings) are only effective if they are designed to be not only audible, but well above ambient background noise levels. As discussed in Chapter 3, aging is accompanied by reduced hearing abilities. Therefore, auditory displays, and alarms in particular, need to be considerably louder to accommodate older ears. Even for navigational guidance, which is less time-critical than collision warnings, straining to hear auditory information takes away cognitive resources that could and should be used to maintain safe driving performance.

Easily adjustable loudness levels for navigational guidance systems enable older adults to adjust the volume or intensity to a level that is comfortable. Older adults will typically need verbal information to be at least 10 dB above a sound level that is adequate for younger drivers. Large individual differences in hearing abilities exist and intensify with advanced age. Therefore, even a 10 dB increase may not be enough for some listeners. Further, the increased prevalence of high-frequency hearing loss and noise-induced hearing loss that can result in reduced ability to hear specific frequency ranges underscores the importance of using sounds that contain multiple frequencies for critical alerts.

6.1.2.1 Verbal messages

Verbal messages have the advantage that they can inform as well as alert the driver. Since language processing is an overlearned task, verbal messages do not require the learning that artificial sounds may require. There is some empirical evidence that over a range of ages, people preferred messages that were solely verbal in nature rather than solely symbolic or a combination of words and symbols.

Verbal messages should be kept terse to avoid overloading working memory resources. Navigational guidance should contain no more than two or three informational elements at a time. This recommendation is consistent with that based on studies of variable message signs for road construction areas (Chapter 5). If complex messages are necessary to provide context for a rapid series of maneuvers, it is important to follow up complex messages with terse messages containing the most important and immediate elements. See Table 6.1 for examples of high and low message complexities. Providing a repeat button for complex verbal guidance is also recommended though may be difficult to implement in practice.

Table 6.1 High- and low-complexity navigational messages

High complexity (for context)	Stay in the right lane to make a right turn on Virginia 185, Nutley Street. Follow signs toward Richmond in 2 miles. Then keep left at the fork to merge onto Nutley Street.
Follow with	
Low complexity	Keep right
	Turn right on Nutley
	Keep left
	Merge onto Nutley

6.1.2.2 Auditory alerts and alarms

Auditory alerts and alarms can benefit all drivers but are particularly helpful to older adults, if designed effectively. Auditory alarms must be clearly detectable, meaning they will generally need to be louder to be audible to older adults due to declining hearing abilities (see Chapter 3).

Effective time-critical auditory collision warnings should have (1) high peak-to-total time ratios (>0.7), meaning that each sound burst should be at its maximum intensity for at least 70% of its duration; (2) rapid tempo (interpulse intervals of 125 ms or less and pulse durations of no more than 200 ms); (3) a base frequency between 1000 and 2500 Hz (part of the frequency range where humans hear the best and the range that is most resilient to aging); and (4) at least three harmonics. Collaborative research efforts by Baldwin in conjunction with NHTSA (Lewis et. al., 2017) led to empirical studies to test perception of crash warnings in different types of simulated vehicles and in an on-road study. Alerts composed of the four parameters described were quickly identified as time-critical warning sounds in all cases, even when presented at 70 dBA against a background of music playing at 75 dBA in the on-road study.

6.1.2.3 Vibrotactile

Vibrotactile DVI displays present information through the sense of touch. Experientially, they may resemble rumble strips along the side of the road or at intersections. But instead of being present on the road surface, they are presented inside the vehicle through vibrotactors, similar to the mechanisms commonly used in mobile phones to present a vibration when the phone is in silent mode.

Physical vibrations or pulses may be presented through the steering wheel or seat cushion, or resistance or force may be applied to the seatbelt or accelerator pedal. Vibrotactile displays have shown promise for use in presenting lane departure and collision warnings to drivers. Most of this work has been conducted with younger drivers though, and there is reason to believe they may not be as effective among older drivers. Tactile sensitivity decreases with age much like the visual and auditory sensory declines discussed in Chapter 3. Decreased tactile sensitivity results in elevated thresholds (the intensity of vibrations must be higher) and decreased discernment of the location and pattern of vibrations.

6.1.3 Multimodal

Multimodal displays tend to benefit all drivers but can be particularly beneficial for older drivers who may have differential impairment in sensory capabilities. For example, for collision alerts, the auditory modality should definitely be used, but also providing a simultaneous visual or vibrotactile alert can increase the probability that the alarm is detected over ambient background noise. Use of multimodal alerts can speed response time, as discussed later. Many affordable new vehicles now have forward collision warnings that use a combination of visual and auditory alerts.

Beyond ensuring that an alert is detectable by an older individual who has differential impairment in at least one sensory modality, multimodal alerts may have particular benefit in speeding response time. As discussed in Chapter 3, age-related slowing of information processing is ubiquitous. However, multisensory integration is enhanced in older adults. This means that older adults benefit more than their younger counterparts when alerts or signals are presented in both visual and auditory modalities concurrently. For example, Laurienti and colleagues (2006) found that presenting stimuli in multiple modalities speeded response time in both younger and older adults, but this performance gain was particularly striking for older participants. In fact, despite being slower than younger participants to respond when stimuli were presented in either visual or auditory modalities, when stimuli were presented bimodally (both visual and auditory simultaneously). older adults were able to respond as fast as the young people could respond in their best single modality.

Adding auditory guidance to visual displays can facilitate comprehension. Traditionally most navigation guidance systems relied primarily on visual interfaces. However, as navigational aids became more common, most included optional voice guidance systems. Early research in this area indicated that older adults prefer to use the voice guidance systems. This allows them to keep their eyes on the road and decreases the switch time between the visual display and the roadway. An Insurance Institute for Highway Safety (IIHS) study (Reagan et al., 2017) concluded from a review of the literature that effects of age on ability to interact with an interface were reduced or eliminated when commands could be issued via a voice interface compared to a visual interface. Most current smartphone-based navigation systems allow the driver to choose either visual-only guidance or combined visual and auditory guidance. If an auditory message was missed, the driver can also look at the written directions. Moreover, the visual guidance can be map-based or verbal (a list of directions).

However, as noted, it is important that the voice guidance be of good quality. Initially, voice navigation systems were of poor quality, such that it was sometimes hard to discriminate between the spoken words "left" and "right." Even at present, there are oddities in pronunciation. For example, one smartphone-based navigation system pronounces the abbreviation for the state of Pennsylvania (PA) as "pa" (rhymes with "ah") while referring to the abbreviation for the state of New Jersey (NJ) by pronouncing the name of each letter separately. This is relevant in light of the long-standing evidence that older people are impaired in what is termed "speech-in-noise," with older people having particular difficulty understanding speech in a noisy environment. Since vehicles provide a noisy environment, this makes the quality of the speech produced in voice guidance navigation systems particularly important for older drivers using such systems. As mentioned in Chapter 8, specific types of training have been shown to improve speech-in-noise perception (Anderson et al., 2013).

6.2 Complexity

Complexity in displays is important regardless of the display modality. Complex messages increase cognitive demand. Complexity can be increased by long or multicomponent messages (e.g., providing several successive turn commands in a navigational aid) or by presenting multiple messages from different systems that all compete for the drivers' limited attention. As more systems are introduced into the vehicle, message complexity can reduce the interpretability of information. The complexity issue is particularly important to avoid exceeding the older drivers' attentional capacity. Messages from multiple systems should be designed in an integrated hierarchical manner so that lower-priority messages (e.g., vehicle status alerts indicating low tire pressure or fuel) do not override and potentially mask more time-critical messages (e.g., lane departure or forward collision warnings).

Visual complexity increases visual search time and strains cognitive processing resources. Complex auditory messages overload working memory processes, increase cognitive workload, and take cognitive resources away from the main task of driving. Avoid long or overly complex messages in both the visual and auditory modalities. Examples of an overly complex visual display are provided in Figure 6.4. The issue of message complexity is addressed in the FHWA's *Handbook for Designing Roadways for the Aging Population*, which recommends no more than one "unit" of information per sign or per phase of sign (for variable message signs). A unit of information refers to a word or phrase that answers one question (e.g., What happened? Where? What should the driver do?). There are also guidelines on the length of messages recommended for work

Figure 6.4 Example of overly complex road sign. (Adapted from https://commons. wikimedia.org/wiki/File:Confusing_street_signs.jpg)

zones, where messages are novel and must be remembered for a short time. Message length is important, because older people have reduced working memory capacity. In California, in work zone "action" words like "Right" and "Closed" can appear in a larger font than other words on a sign. The goal is to make the sign easier to read without enlarging the entire size of the sign, thereby increasing the expense.

Additionally, older adults can be confused or distracted more easily than younger drivers by the presence of many different types of alerts within a given environment. Therefore, the number of different system states that provide auditory alerts should be kept to a minimum, and each one should be clearly distinguishable from the others. This will be a challenge as more automated safety systems and associated icons are added to vehicles.

6.3 Safety systems

Advances in automotive technologies hold great promise for improving the safety and enjoyability of driving. As discussed in the following sections, these come in the forms of both passive and active safety systems.

6.3.1 Passive safety systems

Passive safety systems do not prevent crashes but are features designed to reduce the effects of crashes. They include things like seat belts and airbags as well as the structural design of the vehicle (e.g., safety cages and crumple zones in the front of vehicles).

Seat belts. In a recent year, about half of the people who were killed in crashes were not wearing seat belts. Further, seat belts reduce injuries. Seat belts used by front seat drivers can reduce injuries by 50% in cars and by 65% in light trucks.

Airbags. According to the IIHS, airbags also reduce driver fatalities. However, airbags were designed to be used with seat belts and can injure people not wearing seat belts. The benefit of airbags varies somewhat with its position, but the range is from 29% to 52% reduction in driver fatalities attributable to deployment of airbags. There are also passive safety systems that are designed to help avoid a collision, such as antilock braking systems (ABSs). ABSs have been standard on all new cars since 2000.

6.3.2 Active safety systems

Active safety systems are designed to prevent crashes. They include things like electronic stability control, automatic emergency braking, forward collision avoidance alerts, and blind-spot detection systems. Some of these are mentioned in Chapter 5 and described in detail in Chapter 11. Active safety systems are designed to understand the state of the vehicle and its environment and either to provide feedback to the driver on that state so that a decision can be made by the driver or to automatically initiate some form of vehicle control in an effort to avoid an unsafe event. Many new cars currently have a number of sensors that can detect the potential for forward and rear collisions and provide the driver with a warning in visual, auditory, or vibrotactile format to help the driver guide the control of the vehicle. These sensors essentially function as another set of eyes on the surrounding roadway and can greatly assist older drivers if designed appropriately.

6.3.3 Collision avoidance systems

In light of the greater mortality of older people in crashes, collision avoidance is very important. Typically these systems do not require any input from the driver, except to avoid turning off the system. Early versions provided only auditory and visual warnings, but increasingly common are systems that brake automatically if a collision is imminent. This automatic emergency braking technology has been found by the IIHS to reduce rear-end crashes by about 40% (Cicchino, 2017). Automatic emergency braking is such an important system that 20 automakers have agreed to make it standard on new vehicles by 2022.

6.3.4 Blind-spot indicators

Older adults tend to have reduced range of motion making it more difficult for them to turn their head to see vehicles that may be in their blind spot.

Therefore, as discussed in Chapter 5, blind-spot indicators near the side-view mirrors can be particularly helpful for older drivers. These typically provide a visual alert when a vehicle is close to or is in the driver's blind spot, and some systems also provide an auditory alert if the driver starts to change lanes toward a vehicle in the blind spot to facilitate collision avoidance. The addition of an auditory alert will be particularly important for older people, as the reduced useful field of view (UFOV) characteristic of older people (see Chapter 3) could reduce the ability to detect the signals from the blind-spot detection system.

6.3.5 Rearview backup assist

Restricted range of motion also means that backup cameras and displays can particularly benefit older adults. Beginning in 2018, backup cameras are required in all new vehicles.

6.3.6 Parking assist (proximity sensors)

Restricted range of motion makes parking challenging for older people. Automated systems that assist with parking range from proximity sensors that show how close your vehicle is to vehicles in front and behind to systems that fully park the vehicle automatically.

6.4 Driver support systems

A number of ADASs are increasingly common in modern vehicles and have strong potential for assisting older drivers. ADASs are covered in Chapter 11; therefore, only general design guidance applicable to many of these systems is presented here.

6.5 Displays

6.5.1 Menu structure

Displays of features are generally compiled in menus in one of several formats. Two commonly used types are checkerboard and hierarchical list layouts. Li and colleagues (2017) observed that for younger drivers (aged 22–32 years), checkerboard layouts resulted in safer driving performance at higher speeds (60–80 km/h), while hierarchical layouts resulted in greater efficiency at lower speeds (0 and 30 km/h). It is not clear if this finding holds for older adults.

6.6 Controls

Controls for nondriving essential tasks (e.g., radio tuning, seat position, and climate control) are important to the extent that they minimize the time the driver has to attend to them. Such controls generally require either physical manipulation or voice-speech commands. Direct controls are those where the operator manually manipulates the display (e.g., touch screen displays), whereas indirect controls require movement in one location to control a remote indicator. Examples of indirect remote controls would be a radio-tuning knob that moves an indicator located somewhere else to scroll through stations or menu items. Touch-screen controls reduce the visual demands of the task over indirect controls, where the operator must look at the control and then visually shift to the information display to see if the desired input has been achieved. The more remotely located the control is from the display, the higher is the visual switch time cost. Since, as discussed in Chapter 3, older adults take longer to switch their visual attention from one location to another and then longer to acquire and process the visual information at each location, touch screens can benefit older users considerably.

However, placement and size of touch-screen controls need to take into account the reduced range of motion, physical strength, and psychomotor abilities of older adults. Touch screens that are located too high on the center console, thus requiring the driver to make hand movements to the far right, may be difficult for older adults to reach and manipulate. Interface icons need to be of sufficient size to allow older fingers to make contact without mistakenly activating nearby icons. In a simulator study, Kim and colleagues (2014) found that making touch-screen icons at least 17.5 mm in height and diameter reduced the visual demand of interacting with a secondary task and improved vehicle control. Until research is conducted with a sample of older adults, it is reasonable to assume that for older adults, visual touch-screen icons should be at least this size.

Gesture controls, either through swiping on a touch screen or requiring no actual contact with the actual system but rather hand gestures made in front of a sensor, are increasingly making their way into the DVI space. It is difficult to tell at present the impact these systems will have on safety, as there has been little research to date. Some early investigations, primarily with young operators, indicate that these systems tend to make a number of errors in recognizing the drivers' intended gesture movements, which then need to be erased and corrected. These reliability issues may decrease with extended use of the system. However, frequent errors tend to increase visual distraction since drivers tend to look at the display in order to verify their inputs and to detect and correct errors. This problem may be intensified among older adults who have less precise psychomotor abilities and take longer to learn new motor movements.

6.7 Future in-vehicle technologies

Technological innovations are occurring at a rapid pace, not only in vehicle and physiological sensor capabilities but also in the algorithms that run and integrate them. The sensors and computing capabilities that enable adaptive cruise control, active lane centering, and autopilot features are among the innovations seen in vehicles. Concurrently, sensors capable of tracking heart rate and heart rate variability, eye movements, and other physiological metrics are improving in sensitivity and reducing in cost. In the future these two areas of technological capabilities will likely be integrated to facilitate advanced forms of communication between the vehicle and driver. The following vignette illustrates how this may benefit older drivers in particular.

Doran is 83 years old. He is professor emeritus and a retired, but active, scientist for the Air Force. In fact, he still commutes over an hour each way from his home to his laboratory on base at least 4 days a week. Over the past 10 years he has had periodic heart problems, even being hospitalized on one occasion after suffering a heart attack. Doctors were able to put two stents in, and this had largely, but not completely kept his heart regulated. He still had to actively watch his diet, exercise, and stress levels.

On this particular Wednesday afternoon, he is driving on his usual commute home after a particularly stressful day in the lab. Suddenly an audible alert sounds, and a message pops up on his car's digital display just above the center stack. It reads, "Driver-Monitoring System has detected irregular heart pattern. Please pull over, check your status, and respond with an update." Upon the car sensing no response from Doran, coupled with an erratic steering pattern and the absence of input to the accelerator or brake, the car both switches into automated driving mode and issues an audible alert with the message, "Auto-Driving Mode initiated. Press control sequence to Acknowledge or Disengage." Again, after receiving no input from Doran coupled with an irregular heart pattern, the Autonomous driving mode initiates planned controlled guidance to stop the vehicle at the next available safe stopping point and automatically dials 911 with an emergency message. Again, the car prompts the driver for an acknowledgement and response. But, receiving none, it alerts the 911 operator to the potential medical emergency, the precise location of the driver and vehicle, and then continues to try to obtain a response from Doran. After receiving no response from Doran for another 3 minutes, the system selects from the preassigned list of emergency contacts and informs Doran's wife of the situation and provides the address of the closest emergency medical center. An ambulance arrives within a few minutes and is able to provide emergency resuscitation to Doran and then transports him to the nearby hospital. His wife arrives shortly afterward and is relieved to hear that Doran is stable after suffering another heart attack. With a prescription for statins and orders to change his diet and get more exercise, Doran and his wife are home and thankful for the lifesaving sensor technologies in their recently purchased automobile.

At present, this scenario sounds like science fiction. But, it represents some of the capabilities that future automobiles are likely to have. Advanced automation and specific ADASs are discussed in more detail in Chapter 11.

6.8 Design recommendations

- Use human factors best practices for visual display design, making use of color-coding and Gestalt grouping principles such as proximity and similarity to indicate functional groups of controls and displays.
- Ensure adequate spatial separation of discreet controls to avoid inadvertent engagement or disengagement of separate systems.
- Use standard symbols that fit the users' mental model of systems to promote understanding and avoid confusion.
- Allow the driver to choose the modality in which displays present information.
- Provide the driver with the capability of adjusting the loudness or intensity of auditory displays to ensure detectability and comprehension without taxing cognitive resources.
- Make use of multimodal displays when possible to help offset the increased probability of differential sensory impairment.
- Avoid overly complex messages by limiting message elements to two or three items at a time.
- Provide displays to enhance sensory perception of areas that are difficult to see by drivers with restricted range of motion and UFOV (e.g., blind-spot indicators and rearview cameras).

Recommended readings

Baldwin, C. L. 2002. Designing in-vehicle technologies for older drivers: Application of sensory-cognitive interaction theory. *Theoretical Issues in Ergonomics Science*, 3(4), 307–329.

Harvey, C. 2013. *Usability Evaluation for In-Vehicle Systems*. Boca Raton, FL: CRC Press/Taylor & Francis Group.

chapter seven

Older pedestrians and cyclists

Jerry is 73 years old, an avid cyclist, and a former triathlon competitor. Jerry lives in a suburb about 12 miles from a large metropolis, where his daughter and grandchildren live. Jerry and his wife, Helen, who is 64 and a nurse at the local hospital, only have one car. Jerry is retired, and their budget does not have room for two vehicles. This is not a problem for Jerry who is perfectly happy using his bicycle to get himself around their community. Jerry often cycles around his neighborhood, to the local country store, and on the trail that connects his city to a rural town about 30 miles away. Jerry rarely has any problems while cycling around his community. There are no bicycle paths or sidewalks, but the streets are very wide and well maintained. He is familiar with his neighborhood and the surrounding areas and has plenty of experience and ample space to avoid the danger of being "doored" (hit when a driver opens the door of a parked vehicle into the path of an oncoming cyclist). The local hiker/biker trail is similarly easy for Jerry to navigate. When the weather is nice the trail can become quite crowded with other cyclists and joggers, requiring Jerry to move very slowly downhill with little space to pick up any speed for the next incline. Jerry and Helen's daughter, Tonya, is a waitress and a single mom. Jerry and Helen help out when they can, supervising their 3- and 5-year-old grandchildren. Nevertheless, Tonya's shifts are often unpredictable and conflict with Helen's nursing shifts. Jerry is a great babysitter, but when Tonya is called in and Helen is already at work, his only means of getting to his grandchildren is by his bike. This presents a little more of a problem for him. The distance is not a problem, he routinely cycles 30–40 miles on the weekends, and Helen can pick him and the bike up after her shift, so he only needs to go the 12 miles there. However, once he has left their small community, the roads become larger, with inconsistent numbers of lanes, sidewalks, and paths that are poorly maintained or nonexistent, bike lanes that end without warning, and traffic patterns that cause drivers to need to merge across the bike lane to get into their right turning lane. Twice, Jerry has had close calls, having to swerve out of the way of a car whose driver neglected to look before merging. Once a driver cut him off deliberately, yelling "get off the road" before speeding away.

Jerry's sister, Jenny, 77, has never been as interested as her brother in cycling. Jenny lives alone with her aging yellow Labrador dog, Murphy, in the same neighborhood as her brother Jerry and his wife Helen. Last year,

Jenny suffered a mild stroke that affected her balance and caused her to cease driving altogether. In the past year, the only thing that has forced her to leave her house is when Murphy begs her to take him for walks. This task is daunting for her. The wide streets in the neighborhood, which help Jerry to avoid parked cars, rarely have crosswalks. Neither Jenny nor her dog Murphy can walk quickly anymore, so they take a long time to cross the street. That makes both of them nervous. There are no sidewalks in Jenny's neighborhood, though there are sidewalks once they reach the main road. When cars are parked on the side of the road in front of her neighbors' houses, Jenny and Murphy must either walk on the road, as close to the cars as possible to avoid the traffic, or walk in the gravel and dirt on the side of the road. Murphy has no problem with this choice, but the uneven ground is very difficult for Jenny to navigate. Once they have reached the main road, the sidewalks are nicely even, and crosswalks are plentiful, but there are few choices or possible places for Jenny to stop for a moment to regain her balance or rest. The main road is relatively congested, and next to the sidewalk are thick hedges, so when Jenny stops for a few minutes, she and Murphy block most of the sidewalk. Luckily, Jenny and Murphy have plenty of friends in town, and on the nice days they sit outside at a small restaurant owned by a childhood friend of Jenny and Jerry's. While they sit outside, Murphy napping in the sun, Jenny wonders if she will ever have the courage to walk the 15 minutes here without Murphy. She fears that without Murphy, she may never leave the house alone again.

This vignette illustrates some of the problems faced by older people as they attempt to move through the world outside of the protection of a vehicle. All people are pedestrians at one time or another, so the problems of being a pedestrian are universal. Yet, as we discuss in this chapter, older pedestrians and cyclists are at greater risk than their young and middle-aged counterparts due to reduced health and fitness. Importantly, reduced mobility can influence both physical and mental health. Adults who must rely on others for transport have higher incidences of depression. Loneliness and social isolation are well-established risk factors for increased mortality and Alzheimer's disease. Reduced mobility can contribute to loneliness and isolation. There is increasing evidence that remaining physically active is very important for the health of older people. Physical activity has health-related benefits such as improved mood, balance, agility, and cardiac and respiratory health. This evidence underlines the importance of providing safe and healthy environments for walking and cycling and ensuring that walking and cycling remain viable transportation options for older adults, especially those unable to operate motor vehicles.

Physical fitness in older people is increasingly recognized as important for successful aging. There is recent evidence that modifiable lifestyle factors (obesity-related diabetes and hypertension, physical inactivity, smoking, depression, and educational attainment) account for about one-third of Alzheimer's disease cases in Europe, the United Kingdom, and the United States (Norton et al., 2014). This recent recognition of the importance of lifestyle in Alzheimer's disease risk represents a sea-change in understanding factors governing the quality and duration of life in old age. Both walking and cycling are accessible and low-cost forms of exercise shown to improve health even late in life. Moreover, walking and bicycling also benefit society by increasing population fitness, potentially reducing health-care costs, and decreasing air pollution and traffic. However, it should be acknowledged that people who walk and cycle are exposed to risk of injury and to air pollution. Addressing that question, de Hartog et al. (2010) calculated the costs and benefits of bicycling instead of driving for short trips daily. The benefits of increased physical activity were larger (3–14 months of life gained) compared to the costs of air pollution inhaled (0.8–40 days lost) and traffic crashes (5–9 days lost). Considered together, this evidence emphasizes the overall benefits to individuals and their communities from developing and maintaining infrastructure allowing safe walking and cycling.

Many of the design features that will assist older pedestrians will also assist older cyclists (e.g., improved infrastructure design, signage, etc.). However, there are enough issues that are unique to these two modes of transportation that we address each separately in the sections that follow.

7.1 Pedestrians

In the United States, adults over 70 years of age are the segment of the population with the highest pedestrian fatality rate, as illustrated in Figure 7.1. This problem is not limited to the United States. Older adults are also overrepresented in pedestrian fatalities in many other countries, relative to their younger counterparts (e.g., Denmark, Australia, New Zealand, and France). In Denmark, pedestrians over the age of 70 have an injury rate almost four times higher than that of middle-aged pedestrians. Yet, these fatalities are not inevitable. A recent review concluded that modification of infrastructure could substantially reduce pedestrian-vehicle crashes (Retting et al., 2003).

7.1.1 Patterns of collisions involving older pedestrians

Beyond the sheer numbers of injuries and fatalities in pedestrian crashes accounted for by older adults, the patterns of collisions in which older pedestrians are involved are different from the patterns seen in younger pedestrians. Collisions involving older pedestrians are more likely to occur in daylight hours and are more likely in urban environments, whereas younger pedestrians are more typically struck at night. Older pedestrians are particularly more likely to fall victim: in complex road environments,

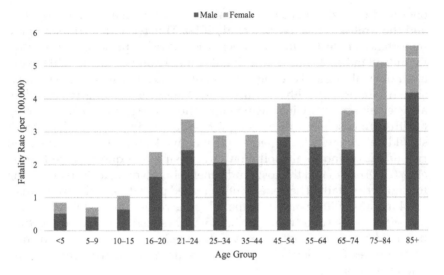

Figure 7.1 Fatality rate (per 100,000) by age group and gender between 2008 and 2012. (Based on NHTSA Pedestrian Safety data.)

including dense or high-speed traffic that must be assessed from multiple directions; in areas lacking controlled, signalized crossings; or in areas where vehicles are often reversing, such as parking lots.

7.1.2 Misperceptions in safe crossing

In general, older pedestrians cross roadways more slowly than do their younger counterparts. This is due in large part to their overall slower walking speeds. Further, older adults are typically less adept at judging safe crossing times than are younger adults. For example, older pedestrians are more likely to cross roadways when there are unsafe gaps in moving traffic. This issue is exacerbated when crossing roadways with higher vehicle operating speeds when gaps are shorter and judging the speed of oncoming vehicles is more difficult. Older adults have been found to use a fixed crossing schema regarding gap judgment and walking speed, whether traffic is low speed or high speed. In contrast, younger adults vary their crossing decisions based on speed of traffic.

Related to misjudgments of appropriate gaps in traffic before crossing a street, older adults also tend to underestimate the time they will actually be in the path of oncoming traffic. This issue is of particular importance in areas where there are multiple lanes of traffic in both directions. Maximizing the safety of the crossing environment with infrastructure treatments (described in the following sections) can occur without the need to educate or change behavior in the millions of current pedestrians and drivers. Further, infrastructure designs that maximize safety for older adults will improve safety for all ages, including children and all individuals with mobility limitations. Therefore, considerable focus is put on infrastructure treatments in the following sections.

7.1.3 Health and fitness

Older adults use roadways differentially based on their health. Where older adults have reported crossing roadways at points other than defined crossing points, those in good health report doing so because they judge it will be faster than going up to a corner, whereas those in poor health report doing so to avoid additional walking and fatigue. It therefore becomes important to ensure that roadways have defined and protected crossings at regular and reasonable intervals to discourage crossing outside of a crosswalk. Alternatively, where it is not feasible or preferable to include defined crossings, benches could be provided for comfortable and protected rest, allowing older adults in poor health to rest frequently and make better crossing choices.

Older pedestrians may not always find elevated overpass walkways easy to use. These elevated walkways are designed to allow the safe crossing of major roadways but include steep ramps and/or stairs. Older pedestrians may be unable to climb long flights of stairs due to low physical fitness, low respiratory capacity, joint, and balance problems. Where the use of pedestrian overpasses is unavoidable or necessary, it is important to clearly indicate alternatives, such as elevators or alternate route suggestions.

7.1.4 Personal safety

Personal safety is a particular concern of older adults. Older adults are typically less agile and frailer than are younger adults and therefore less able to physically fend off an assault. Older adults are concerned about walking in metropolitan areas due to fears of attack by people or dogs as well as fears of being struck by a vehicle and of falling. Safety from crime has been reported as a major concern for older people. Living in a neighborhood that is viewed as unsafe at night has been found to reduce regular physical activity among individuals, especially women. Ensuring that public walkways include adequate lighting is one way to improve perceived risk and safety.

7.1.5 Infrastructure treatments

Infrastructure can play an important role in pedestrian safety, especially for older adults. The US Federal Highway Administration's (FHWA)

Handbook for Designing Roadways for the Aging Population provides recommendations for road designs for older people. The recommendations include walking speed assumptions, crossing signals, crosswalks, educational signs, "vehicles yield to pedestrians" signs, and specifications for refuge islands. We discuss some of the better supported recommendations in this section and refer the reader to the FHWA's handbook.

7.1.5.1 Personal safety

Fears of older people for their personal safety while walking can be assuaged by sidewalks that are not bordered by tall walls and shrubbery that blocks sightlines. Where walkway underpasses are unavoidable, they should be clean and well maintained. It is advisable to include 24-hour lighting, even in short tunnels. Law enforcement surveillance cameras could be used to offer additional peace of mind to pedestrians.

7.1.5.2 Crossing signals

Crossing safely at a signalized intersection is only possible when the time allotted by the signal is greater than the time it actually takes a pedestrian to cross the roadway. Older adults have slower walking speeds than the population as a whole. In 2009, the FHWA lowered the allowed walking speed from 4.0 to 3.5 feet per second, with the option to slow that further to 3.0 feet per second when a street is very wide or other circumstances dictate the need for more time for crossing.

One design solution that was specified in the 2009 FHWA's *Manual on Uniform Traffic Control Devices* (MUTCD) is the addition of a countdown to crossing signals at signalized intersections. A countdown signal is now required for intersections when the pedestrian change interval (flashing upraised hand) is 7 seconds or greater. Figure 7.2 shows a countdown crossing signal. This feature replaces the flashing "walking person," which apparently does not convey meaning to most people.

The countdown is informative in that it tells pedestrians both whether they can cross and the amount of time remaining to cross the roadway. A steady upraised hand indicates that a pedestrian should not enter the intersection. There is now the option of an addition of "animated eyes" to the top of the display. These are white outline eyes with white eyeballs that move from left to right, reminding pedestrians to watch for vehicles in the intersection. These animated eyes may be particularly useful to older people as a reminder to be vigilant for traffic. Sounds are also optional and can be messages or tones audible 6–12 feet from the display of the signal. Sounds are important for people with low vision, more common among older people.

7.1.5.3 Crosswalks

Crosswalks are the most commonly used pedestrian protection. Although the MUTCD sets standards for crosswalk markings, some states, towns, and

Figure 7.2 Countdown crossing signal required by 2009 MUTCD for intersections when the pedestrian change interval (flashing upraised hand) is 7 seconds or greater. (This file is licensed under the Creative Commons Attribution-Share Alike 3.0 Unported license.)

other jurisdictions choose to develop their own standards so that there is a lack of uniformity. For crosswalks, the handbook recommends markings that resemble ladders (Figure 7.3) on the basis of evidence that they are more visible. Such high visibility crosswalks have been designated by the FHWA as a "promising practice." They can help both the older pedestrian and the older driver with reduced acuity and contrast sensitivity to see the crosswalks.

Common sense would suggest that pavement markings at crosswalks increase pedestrian safety, but under some conditions the opposite seems to be the case, particularly for older adults. The FHWA summarizes evidence that on busy, multilane roads, pedestrian strike incidents occur more often at marked compared to unmarked crosswalks (defined in certain

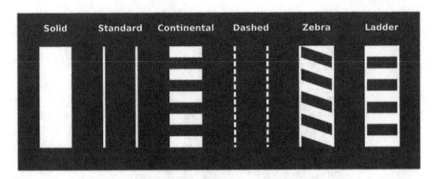

Figure 7.3 Different styles of crosswalk, with the "ladder" crosswalk recommended by the *Handbook for Designing Roadways for the Aging Population* at the far right.

states to exist in any intersection regardless of markings). This effect is even greater for older adults at intersections without stop signs or control signals (Koepsell et al., 2002). On two-lane and multilane roads with low traffic volume, whether crosswalks are marked or unmarked has no effect. However, most of the pedestrians crossing multilane roads are doing so in marked crosswalks, particularly young and older pedestrians. There is also evidence that the difference in pedestrian strikes between marked versus unmarked crosswalks is due to "multiple-threat" crashes. These crashes occur when a motorist in one lane stops to yield to a pedestrian, but a motorist in an adjacent lane does not see the pedestrian crossing in front of the adjacent stopped vehicle and strikes the pedestrian. Refuge islands (described later) are a method of alleviating this issue when the threat comes from the opposite direction. Pedestrian hybrid beacons (described later) are another way to make crossing safer at nonsignalized intersections where about 75% of pedestrian fatalities occur. Better nighttime lighting overall increases the detection of pedestrians in crosswalks and ensures that particularly older pedestrians can easily see the walkway. For specifics, see the *Handbook for Designing Roadways for the Aging Population*.

As mentioned briefly in Chapter 4, there are also technological approaches to reducing risk to pedestrians. Some vehicle manufacturers have developed "pedestrian detection" systems that use a camera and radar to attempt to detect people (and bicyclists for some manufacturers) in the vehicle's path. These systems are designed to warn the driver and in some systems to automatically apply the brakes if needed. These systems do have their limits, which are detailed in owner's manuals. For example, systems may not recognize very short or very tall pedestrians, those carrying a large object, or pedestrians at dawn or dusk. These systems may not completely avoid a collision but could mitigate one. NHTSA researchers analyzed the effectiveness of pedestrian detection systems in the context of historical crash data and estimated such systems would prevent 5,000 vehicle-pedestrian crashes each year (810 fatal crashes).

7.1.5.3.1 Sidewalks Sidewalks should be wide enough for older adults to walk using physical supports, such as walkers or canes, or supported by a walking partner with space for other pedestrians to pass easily without jostling. Sidewalks should be well maintained, kept even, and resurfaced whenever possible to avoid uneven cracks or other fall hazards. The 2010 standards of the Americans with Disabilities Act (ADA) require "curb ramps or other sloped areas" at intersections that have barriers to entry from the street. These ramps are required for streets and roads that have been newly constructed or newly altered (e.g., curbs). A link to the current ADA standards is provided at the end of this chapter. Sidewalks that have been damaged by weather or tree roots, as well as those with excessive debris such as leaves, pose a hazard to older pedestrians. Uneven pavements with cracks can increase the risk of falls, especially for older pedestrians with impaired balance.

7.1.5.4 Pedestrian refuge islands

Refuge islands are defined areas in the median of a road that provide a place for pedestrians to stand with reduced risk of being struck by passing vehicles (Figure 7.4). Often refuge islands are on raised curb medians. Refuge islands are important for older pedestrians whose slower walking speed means they may not be able to cross the full width of a road during one light cycle.

Figure 7.4 Refuge island. (This file is licensed under the Creative Commons Attribution-Share Alike 3.0 Unported license.)

7.1.5.5 Road design treatments

There are recent innovations in road design that can assist pedestrians directly and indirectly. The FHWA has highlighted a number of "proven safety countermeasures." One of these is the pedestrian hybrid beacon designed to help pedestrians cross safely in the middle of a block or at unsignalized intersections. The beacon has two red lenses arrayed below one yellow lens (Figure 7.5) and is activated by a call button on the sidewalk. The beacon is normally dark. When a pedestrian pushes the call button, the beacon starts a sequence of flashing yellow to red lights with eventually the red light turning steadily on. At that point, the pedestrian gets a "walk" signal. Once the pedestrian has crossed, the signal again goes dark.

Another "proven safety countermeasure" is the "road diet," which typically converts a four-lane, undivided road with no refuge islands, to a road with two through lanes and one two-way left-turn lane in the center (Figures 7.6 and 7.7). The road diet reduces speed overall, reduces rear-end and left-turn crashes due to the center turn lane, and provides a place for pedestrian refuge islands. "Curb-extensions" (Figure 7.8) extend the sidewalk into the street at the crosswalk to improve pedestrian safety. Both road diets and curb extensions increase safety for pedestrians by shortening time needed to cross a road. These structural changes also speed traffic by allowing shorter pedestrian-only signals. They have been shown to reduce overall crash frequency, especially in smaller cities.

7.2 Bicyclists

Cycling holds great potential as both an enjoyable way of improving physical fitness and a means of transportation. The challenges faced by

Figure 7.5 Pedestrian hybrid beacon. (By Michael Barera, CC BY-SA 4.0, This file is licensed under the Creative Commons Attribution-Share Alike 4.0 International license.)

Figure 7.6 Example of a road diet transformation, before the transformation. (This file is licensed under the Creative Commons Attribution-Share Alike 3.0 Unported license. By TransportObserver – My own camera, CC BY-SA 3.0, https://commons. wikimedia.org/w/index.php?curid=39583284)

Figure 7.7 Example of a road diet transformation, after the transformation. (This file is licensed under the Creative Commons Attribution-Share Alike 3.0 Unported license. By TransportObserver – My own camera, CC BY-SA 3.0, https://commons. wikimedia.org/w/index.php?curid=39583284)

the older cyclist are similar in many ways to those faced by the older pedestrian. As discussed in Chapter 3, older adults generally have slowed processing times, and therefore may have difficulty quickly assessing their surroundings. Would a swerve move the cyclist into faster traffic? Would a quick stop risk losing control of the bicycle? These issues can be somewhat alleviated by the use of protected bike lanes, described later.

Figure 7.8 Curb extensions. (Adapted from Arpingstone. The original uploader was Arpingstone at English Wikipedia. Transferred from en.wikipedia to Commons. Public Domain, https://commons.wikimedia.org/w/index.php?curid=3513326)

However, these protected lanes are often nonexistent in rural areas where cyclists must share the road with motor vehicles. Protected bike lanes must be wide enough to allow riders to safely dismount if, for example, they become fatigued or are on a particularly steep hill. The remainder of this chapter will address design issues and guidance related to older cyclists.

7.2.1 Safety and patterns of collisions

Older people increasingly rely on bicycling. In countries like Germany and the Netherlands, about half of all trips made by people over 75 are conducted by walking or bicycling. In the Netherlands, 75% of trips by that age group are made by bicycle. Yet, in the United States in survey after survey, people contemplating bicycling in cities say the biggest obstacle is fear of being struck by a car. Survey research in Portland, Oregon, shows that 60% of people are interested in cycling on a regular basis but concerned about their safety while cycling. Older cyclists, like older pedestrians, do experience a higher risk of injury in a crash. In Denmark, a country where cycling is a prevalent form of both transportation and leisure, adults over the age of 70 have injury rates about three times higher than those of middle-aged adults. These rates appear to be rising. A review of bicycle trauma injuries between 1998 and 2015 found that age-adjusted bicycle-related injuries rose by 28%, while related hospital admissions rose by 120%. Importantly, that increase was driven largely by older people. The proportion of injuries in people over age 45 years increased 81% over that time period. Moreover, the proportion of severe injuries in older people increased over the period. These patterns are consistent with a shift in the demographics of cyclists who are increasingly older. As older people increasingly use technology such as fitness trackers or GPS displays to monitor their cycling speed and routes, there may be particular risks for older people who have known susceptibility to distraction.

7.2.2 Infrastructure design

Designers should also consider infrastructure design from the viewpoint of older cyclists. Older cyclists ride an average of 20% slower than their younger counterparts, and this includes through intersections. This means it takes older cyclists longer to traverse intersections leaving them vulnerable to red light running and at greater risk of intersection crashes. Signalized bike paths such as the one illustrated in Figure 7.9 can assist older and younger cyclists alike, but may be particularly helpful for older cyclists. As older cyclists require additional time to traverse intersections, the amber phases of the light should be a minimum of 3 seconds and preferably longer.

7.2.2.1 Protected bike lanes

Over the years there have been differences of opinion on how to best protect cyclists from cars and trucks. Previously it was argued that cyclists were safest when they behaved like a car and rode in the center of a lane. However, empirical studies looking at the effect of route infrastructure on cyclist injuries found that cyclist safety is enhanced by protected bike lanes, off-street bike paths, and intersections with speed limits below 20 mph. Cyclists experience increased risk in shared car/bicycle lanes and traffic

Figure 7.9 Signalized bike path. (This file is licensed under the Creative Commons Attribution-Share Alike 3.0 Unported license.)

circles. Despite high rates of helmet use in the United States, the bike fatality rate has remained very high compared to countries in western Europe that have seen sharp drops in bike fatalities despite their low rates of helmet use. There are a number of innovations in road treatments for cyclists, but mainstream road engineering organizations (FHWA or AASHTO) have been slow to embrace them. Yet, because AASHTO is considered the road engineering standard, many municipalities will not add protected bike lanes in the absence of published standards from AASHTO. To fill the gap, transportation officials from 15 cities have created the *Urban Bikeways Design Guide* that standardizes innovative treatments for cycling. The *Urban Bikeways Design Guide* is the main source of information for cities who seek better road treatments for cyclists. Bike boxes and protected bike lanes are innovations that are being increasingly used in cities. Bike boxes (also called "advance stop lines") are markings on pavement at signalized road junctions defining a space in front of vehicles that is reserved for bicyclists (Figure 7.10). Bike boxes can be very important in protecting bicyclists from large trucks which have a considerable blind spot in front of the cab.

Bike boxes allow bicyclists to start before motor vehicles when the traffic signal changes from red to green. Bike boxes can also improve pedestrian safety by increasing the physical separation between pedestrians and motor vehicles. Protected bike lanes (called cycle tracks in Europe) are lanes adjacent to a roadway or in a roadway that are distinct from that roadway by means of barriers or elevation. The term does not refer to bicycle lanes within an all-vehicle roadway in which bicyclists are not protected from cars and trucks. Academic studies have found that the presence of protected bike lanes increased bicycle traffic. A number of studies have found that separated bike lanes reduce both death and injury,

Figure 7.10 Bike box. (This file is licensed under the Creative Commons Attribution 2.0 Generic license.)

especially when they have dedicated signal phases. AASHTO's revised guide with needed specifications for bike lanes in the United States is expected in 2018.

7.3 Design recommendations

- Intersections and crosswalks
    - Crosswalks should be clearly marked and should be accompanied by traffic controls where traffic is dense.
    - Intersections should always have designated crossing areas equipped with crossing lights.
    - Crossing signals should include a countdown showing the amount of time left for crossing.
    - Crossing times used in signal design should take into account the slowed walking speeds of older adults.
    - Refuge islands should be present between divided lanes of traffic to allow sanctuary if pedestrians are unable to cross fully in the time allotted.
    - Where pedestrian traffic is very heavy or there are large distances between intersections, midblock crossings should be used to reduce the perceived need to cross without a crosswalk. Pedestrian hybrid beacons would improve safety at such crossings.
    - Curb extensions can reduce crossing times needed.
    - To decrease motorists' speeds at intersections where pedestrians are common, consider converting stop-sign controlled intersections into roundabouts (described in Chapter 5).
- Walkways
    - Avoid overuse of underpasses and overpasses as they often place the burden of crossing on the pedestrian and may have limited accessibility.
    - Where underpasses or tunnels are necessary, ensure they are well maintained and well lit.
    - Consider including surveillance in excessively long or less-travelled underpasses.
    - Where overpasses are necessary, ensure accessibility in the form of an elevator.
    - Avoid the use of stairs where low-incline ramps can be used.
    - When stairs are unavoidable, ensure information is provided about alternate routes.
    - Avoid high walls or shrubberies that block sightlines or create a closed-in feeling.
    - Keep surfaces even and well maintained.
    - Consider including benches and shaded rest areas, particularly for longer stretches of path.

- Bike lanes
  - Consider including protected bike lanes on the sides of roadways or, where appropriate, in parallel with pedestrian paths.
  - When bike lanes are part of the roadway, adding protection (physical separation between cyclists and motorists) increases safety.
  - Keep surfaces even and markings well maintained.
  - Where bike lanes are segregated from the roadway, ensure signal behavior is specified for bicyclists and is safely coordinated with both roadway signals and pedestrian signals. Ensure that the amber signal allows ample time for slower cyclists (e.g., minimum of 3 seconds for an average 100-foot intersection).

Recommended readings

National Association of City Transportation Officials (NATCO). *Urban Bikeway Design Guide*. Retrieved from https://nacto.org/publication/urban-bikeway-design-guide/

2010 Standards of the Americans with Disabilities Act (ADA). Retrieved from https://www.ada.gov/regs2010/2010ADAStandards/2010ADAstandards.htm#curbramps

chapter eight

Design of older adult transportation training programs

The importance of mobility for quality of life is clear. Less clear may be how older adults remain mobile by learning to use new forms of transportation and relearn or improve strategies to maintain safety using their existing forms of transportation. There are numerous older driver and transport training programs in existence. Training on the use of newer forms of automation and advanced driver assistance systems (ADASs) is in need of development. Here the focus is on designing effective training programs. We briefly discuss the evidence, or need for evidence, for key elements of effective programs. Programs that facilitate mobility in older people can result in more older people who are able to age in place, fewer nursing home placements, better quality of life, and a tremendous cost savings for individuals and society as a whole. Design of effective older transportation programs hinges on identifying and implementing the important components that produce improvements that have been confirmed by evidence. At present there is lack of definitive evidence for the effectiveness of many existing programs. However, there is some evidence that skill-specific training and computer-based driving simulation training offers some benefit to older drivers.

Designing effective older-driver training programs depends on adequate understanding of both the age-related factors that contribute to increased crash involvement and the skills and abilities that are amenable to training in older people. Equally important are effective driver assessment techniques to identify those individuals in need of training and to determine whether or not training has been successful. David Eby and Lisa Molnar, two experts in the area of older driver training programs from the University of Michigan Transportation Research Institute (UMTRI), point out that screening and assessment are different domains, and training, though related, is a separate construct itself. Screening, they point out, is part of an overall program and should not be used to make licensing decisions. But, screening may be used to determine whether an individual might benefit from a training program or be referred for further assessment. Assessment, often involving cognitive, medical, and physical testing, is aimed at determining the underlying mechanisms involved in functional impairments and may be the basis for determining the degree of impairment and for making licensing decisions. Here, we focus primarily on screening for risk—determining when an older adult may benefit from a training program and the design of that training program.

There are several approaches to older driver training. One of the most common types includes computer-based driving simulation training. Another approach involves classroom-type knowledge-based training. The comprehensive and effective approaches combine the two into a program that involves classroom-type education, specific skills training, and individualized simulator instruction or on-road instruction.

Training that targets specific abilities, such as physical fitness and speed of processing, shows promise for reducing crash risk among older drivers. It is imperative that older-driver training programs be designed to focus on evidence-based strategies for improving safety. Concurrently, well-designed programs that train older adults how to use alternative forms of transportation will facilitate safe mobility for both people who are currently driving and those who have stopped driving. Designing effective training programs for alternatives to driving is an important aspect of maintaining safe mobility among older populations.

In this chapter, we examine the factors that lead to documented change in ridership among older adults for those programs aimed at public transportation and improvements in crash involvement for those programs aimed at older drivers. The specific focus is on factors that improve the design and thus outcomes of transportation training programs for older adults.

8.1 Essential components of older-driver training programs

Effective older-driver training programs contain a number of different elements designed to target specific cognitive, physiological, self-awareness, knowledge, and experience-based domains. Effective programs target each domain and educate seniors about age-related changes that impact driving, methods of assessing their own capabilities and risk, and alternative methods of transportation when they must cease driving. There are two approaches to older-driver training: information processing and driving skills. The former attempts to fundamentally change the ability to process events; the latter attempts to improve the specific skills used in driving. In general, training programs with some form of individualized feedback and on-road or simulator training components tend to show greater effectiveness than in-class training by itself (e.g., see Sawula et al., 2017). We begin with a discussion of key problem areas, risk awareness dimensions, and their assessment, and then turn to a discussion of components that have evidence of amelioration of deficits with training that can increase the effectiveness of older-driver training programs.

8.2   Key problem areas

Age-related changes in sensory, cognitive, and psychomotor abilities that can impact driving were discussed in Chapter 3. We do not review each of these again here but rather note that throughout this book we have been referring to designs that will improve life for normal functioning older adults. Screening for more severe cognitive impairments and dementias is necessary to identify individuals who may require medical treatment and who would benefit little from standard older-driver training programs.

8.2.1   Cognitive impairment and dementia screening

Before enrolling in a training program, older adults should be free of acute illness, and it should be determined that they possess normal age-related cognitive abilities. One commonly used screening tool to ensure this level of functioning is the Mini-Mental State Examination (MMSE; Folstein et al., 1975). This is a low-cost and easy to administer test that can be administered by a primary care physician to screen for dementia or mild cognitive impairment (Folstein et al., 1975). Failure to achieve a score of 24 or above on the 30-point MMSE would be indicative of potential cognitive impairment and increased risk of both crashes and anosognosia (failure of metacognition and associated failure to be aware of their impairment). Anosognosia is a particular challenge for the older driver, since he or she will be unaware of the degradation of driving abilities and thus will fail to self-regulate driving. For cognitively normal older adults, training programs should include risk awareness.

8.2.2   Risk awareness

One important component of effective training programs is providing a knowledge-based account of how advancing age impacts driving ability, the impact of certain medical disorders and medications on driving, and common problem areas for older drivers. Effective training programs will provide knowledge of these issues to assist older drivers with risk awareness. See Table 8.1 for a list of essential elements of risk awareness.

8.2.3   Rules of the road and traffic safety

There is considerable evidence that older-driver training that promotes improved knowledge of laws and rules of the road is effective based on formal tests of this knowledge. Effective training programs should therefore include this classroom-type instruction. In some cases, older adults may simply need a refresher course of information previously learned. However, there are a number of new laws that may not have been learned when initial licensure was granted and therefore training on this information is beneficial.

Table 8.1 Key risk awareness issues for older drivers

Dimension	Key risk areas	For more information
Cognitive ability	Confusion and memory failures while driving, getting lost, MMSE scores <24—particularly impaired design copying	For data regarding the significance of MMSE and impaired design copying, in particular, see Marottoli et al. (1994)
Visual ability	Severe visual acuity impairment (>20/70 binocular distance), contrast sensitivity, and poor visual attention and UFOV	
Motor ability	Inactivity (walking less than one block per day), three or more foot abnormalities (e.g., toenail abnormalities, bunions, calluses, toe deformations like hammertoe), severe restricted range of motion	For odds ratios for inactivity and foot abnormalities, see Marottoli et al. (1994)
Medication usage	Opioid analgesics, tricyclic or tetracyclic antidepressants, antipsychotic medications, benzodiazepines	See AAA's Roadwise Rx for further information about how specific prescription and over-the-counter medications impact driving: https://seniordriving.aaa.com/understanding-mind-body-changes/medical-conditions-medications/roadwise-rx/
Driving behaviors—risky	Inadequate scanning, problems backing up and making left turns	
Driving history	Recent crashes, moving violations, police stops, low annual miles driven	For low mileage as a risk factor, see Langford et al. (2013)
Medical disorders	Certain medical disorders are known to increase crash risk (e.g., dementia, Parkinson's, metastatic cancer, neurodegenerative disorders)	For detailed information about specific disorders, see FMCSA website: https://www.fmcsa.dot.gov/regulations/medical/reports-how-medical-conditions-impact-driving
	Dizziness, loss of consciousness	
Alcohol and other drug abuse	Intoxicated driving, recent history of documented or self-reported abuse	

Abbreviations: FMCSA, Federal Motor Carrier Safety Administration; MMSE, Mini-Mental Status Examination; UFOV, useful field of view.

8.2.4 Self-assessment

Just as important as the ability to drive safely is accurate self-assessment of one's driving ability. Effective programs should provide opportunities for older adults to rate their driving abilities in important areas and to compare their ratings to those of an independent observer (i.e., an occupational therapist trained in driver assessment). Both younger and older drivers tend to overestimate driving skills. As older drivers' abilities decline, appropriate calibration of their self-assessments and their actual abilities is extremely important. Training programs that provide individualized feedback on performance can assist individuals with accurately calibrating their perceived abilities with their actual abilities. It is important to base such individualized feedback in part on driving situations known to be problematic for older drivers, notably left turns. Appropriate calibration of perceived and actual driving abilities assists the older adults with self-regulation or driving behaviors (e.g., avoiding problematic intersections) and determining when driving cessation is necessary.

Older drivers who do not show evidence of anosognosia generally tend to moderate their own driving. For example, they will often avoid driving at night, during rush hour, or during adverse weather. However, effective older-driver training programs should assist seniors with accurately assessing their driving abilities and in understanding the age-related threats to safe driving.

In the next section we discuss specific training components that are supported by evidence of effectiveness in improving driving performance and reducing the crash risk of older drivers. Keep in mind that this list is not exhaustive, nor is it meant to indicate that the evidence is conclusive. Rather, it is presented to inform best practices based on current research.

8.3 Abilities amenable to rehabilitation/training

8.3.1 Physical and psychomotor training

Improvements in health status and fitness achieved through physical exercise programs have been shown to improve both cognitive and driving performance among older adults. Older adults experiencing limitations in flexibility of the lower body and spine, in particular, appear to benefit in improved driving performance from training programs that increase flexibility and range of motion. Given the beneficial impact of physical fitness training on ameliorating a wide range of negative outcomes, it is surprising that more research has not been conducted on its effectiveness for improving driving performance. The research that has been conducted indicates that aerobic exercise improves overall health, information processing speed, and attention. Further, general physical fitness exercises emphasizing flexibility and range of motion have been found to improve driving behavior.

8.4 Overall fitness

Programs that improve overall fitness may have both direct and indirect benefits on mobility. Improved health and fitness tends to improve cognitive functioning and decrease the prevalence of and risk for other medical conditions. Aerobic exercise improves information processing speed, attention, executive function, and memory (for a critical review of a number of studies see Smith et al., 2010). Fitness programs for older people that target walking (at least 1 hour twice a week) have improved walking mobility and improved oxygen utilization relative to baseline after 1 year (Valenti et al., 2016). Additionally, general exercise programs may also improve overall health as well as flexibility and range of motion.

8.4.1 Range of motion

Exercise programs that improve range of motion among older adults can improve driver safety. Improved range of motion facilitates the ability of older drivers to turn their heads to see vehicles that may be in the blind spot or to detect vehicles, pedestrians, and objects that are obstructing their paths while backing up. For example, Marottolli et al. (2007a) examined the impact of a 12-week randomized exercise conditioning program emphasizing physical abilities thought to be related to driving ability. The exercises included axial/extremity conditioning (e.g., cervical, truck, and shoulder flexion and abduction) as well as upper extremity coordination/dexterity, hand strength exercises, gait improvement, and other physical exercises. The exercises were designed to take 15 minutes per session in total, and participants were asked to complete them once per day. After 12 weeks of the exercise training, improvements were seen in driving performance metrics, relative to drivers' own baseline and a no-exercise randomized control group. Individuals with the lowest driving performance scores at baseline showed the most dramatic improvements, but even higher-scoring individuals showed significant improvement.

8.5 Skills and knowledge amenable to training

8.5.1 Rules of the road

Older driver training programs should include knowledge-based instruction on rules of the road. Several studies have documented improvement in knowledge tests following training even for people who have been driving for a long time.

8.5.2 Cognitive training

A number of cognitive training programs have demonstrated some evidence for their potential to decrease at-fault crash risk among older adults. For example, one large-scale randomized clinical trial found support for both cognitive reasoning and speed of processing training (Ball et al., 2010). The cognitive reasoning training involved learning strategies for solving series letter and word problems. In the speed of processing training, people practiced identifying and locating visual items in increasingly complex displays. Ball et al. found that older adults who engaged in up to 10 sessions of either reasoning or speed of processing training had approximately 50% fewer crashes (per mile driven) than a control group over a 6-year period following training.

8.5.3 Speed of processing training

Speed of processing training has shown considerable promise for reducing the rate of motor vehicle crashes among older adults. Several studies have found that speed of processing training reduces the number of dangerous road maneuvers and appears to decrease crash risk as assessed by motor vehicle records. The most common form of speed of processing training used in older driver training is the useful field of view (UFOV) test (Ball et al., 1988). Once a person's discrimination threshold is determined for events that appear centered in front of the person, other events are simultaneously presented that must be located in space up to 30° of eccentricity in the periphery of vision. That aspect of the task requires divided attention. Finally, irrelevant, distracting events are added to the display. That aspect of the task requires selective attention. Training generally involves multiple sessions with identification, discrimination, and localization tasks at 10 different display durations. The training is conducted in small groups, led by a trainer. Ball et al. conducted 10 sessions of 60–75 minutes each conducted over 6 weeks. Booster training was provided in years 1 and 3. Speed of processing training was associated with better "instrumental activities of daily living" (e.g., personal self-care, managing money, shopping, taking medication, using the telephone) and even reduced risk of dementia in one study.

Important for driving safety, speed of processing training has been shown to reduce unsafe driving maneuvers in healthy older people during real-world driving sessions, with effects lasting 18 months (reviewed in Ross et al., 2016). Ross et al. (2016) found a dose-dependent effect of speed of processing training on driving, meaning that the benefit increased with more training sessions. Participants completed 0–18 sessions, and those who had completed more sessions tended to drive with greater frequency across the 5-year study period, relative to participants in the control group. The authors defined *driving frequency* as the average number of days per week that a person drove. They found that individualized, adaptive speed of processing training sessions had greater benefit than standardized training, and this training had the greatest impact on individuals who had the lowest baseline performance. They reasoned that individuals with poor baseline performance had the greatest room for improvement.

8.6 Driving skill training

Another approach is to train driving skills directly. A number of driving-related cognitive skills show improvement with training though their impact on crash risk is less clear. Among the training is training aimed at hazard perception and training aimed at improving intersection scanning behavior.

8.6.1 Hazard perception training

Hazard perception training is one type of training for which there is mounting evidence of effectiveness in reducing crash risk. Hazard perception training has been implemented in the driver licensing process in the United Kingdom due to the evidence of its effectiveness. That training is generally easy to implement via computer-based programs and could be included in older adult training programs. Hazard perception has been shown to improve among older adults undergoing the video-based training both immediately following training and at 3- and 6-month follow-ups (Horswill et al., 2015).

8.6.2 Intersection scan behaviors

Older drivers are at greater risk of crashes when making left turns at intersections without controlled left-turn lights (discussed in Chapter 4). This appears to be due to scanning behavior. Although older drivers detect road hazards at the same level as young drivers, they are less likely to fixate their gaze on *potential* hazard zones. Importantly, older drivers are less likely than younger drivers to scan *outside* the planned path of the vehicle before completing a left turn. Training programs should not only inform older drivers of their greater crash risk at intersections, but also attempt to train scanning patterns. Fortunately, training programs designed to improve scanning performance at intersections show promise for offsetting this scanning pattern (Marottoli et al., 2007a,b; Pollatsek et al., 2012). Older-driver training programs should include a component of teaching, or reteaching appropriate scan behaviors, particularly at intersections.

8.6.3 Multitasking training

Driving and many other real-world tasks require multitasking. Multitasking requires the ability to exert cognitive control over multiple simultaneous goals—abilities that are demanded by driving at times and frequently found to be compromised by advanced age. Specific training targeting cognitive control in multitasking situations may offset age-related declines in these abilities. For example, Anguera and colleagues (2013) utilized a driving-like video game they called "NeuroRacer" to both assess and train adults on a visuomotor tracking task (e.g., simulated driving) and a perceptual discrimination task (e.g., sign recognition). Initially, they found a linear relationship between multitasking performance and age, such that abilities to perform both tasks at the same time decline with increased age. However, 4 weeks of adaptive multitask training (i.e., the difficulty level progressed based on each individual's current performance) significantly improved performance of adults aged 60–85 years, with the benefit persisting 6 months later. The trained adults reached levels on the task attained by untrained 20-year-olds. It is likely that aspects of controlled attention, which play an essential role in working memory, are at least partially responsible for multitasking improvements.

8.6.4 Cognitive training

There has been considerable attention in recent years to a number of programs aimed at improving cognitive function, and working memory in particular. Numerous commercially available programs and games tout their efficacy. However, there is a lack of consistent evidence of the efficacy of these commercially available training programs from carefully controlled scientific studies. The scientific literature on cognitive training is characterized by active debate concerning its effectiveness. When many of the common confounds stemming from poor research design (i.e., lack of an active control group to rule out things like test-retest improvement, increased social interactions, placebo effect, etc.) are accounted for, there is no consistent evidence of far transfer of training. Performance improvements from training do not tend to improve performance on any task other than the one used for training. Some meta-analyses (analyses conducted on a number of similar studies) assessing effectiveness of cognitive training have found evidence of far transfer in older people with small effect sizes, while other meta-analyses have not. However, as Greenwood and Parasuraman (2016) suggest in a recent review article, there is evidence that cognitive training aimed at improvements in attentional control and distraction suppression (see summary that follows) can lead to durable far transfer effects.

8.6.5 Summary of driving skill training programs

Considerably more research is needed to determine the extent to which physical and cognitive abilities can be retained or regained by educational programs. In general, programs aimed at improving the physical fitness and cognitive fitness of older adults can do little harm and have the potential to improve overall health as well as physical and cognitive abilities. Knowledge-based programs to teach or reteach rules of the road, hazard perception, intersection scan patterns, and better-calibrated self-assessment of driving ability show promise for improving the safety and mobility of older adults. But unfortunately, when put to the test of scientific rigor, working memory training may not offer a magic bullet to offset age-related changes. In contrast, cognitive training that focuses on rapid suppression of distractors and enhancement of targets appears to transfer to untrained tasks, including driving. Specifically, "speed of processing" training, certain forms of working memory training that require distraction suppression, and multitasking training that requires speeded distraction suppression can be good models for cognitive training to enhance driving. As reviewed previously, speed of processing training has been shown to improve driving.

Older driver training programs will be most effective if they are part of an overall community-based approach that involves seniors, their family members, health-care providers, occupational therapists, social service workers, and police officers. The National Highway Traffic Safety Administration (NHTSA) recently published recommendations for Model Programs (Table 8.2).

8.7 Public transport training

Older adults living in urban areas generally have access to various forms of public transportation, like buses and trains. However, novice older users may need training on how to access these services. According to a recent large-scale report sponsored by the Federal Transit Administration (Burkhardt et al., 2014), nearly every transit company with a website offers at least some form of transport training. However, this training ranges widely from a web page providing information for new passengers to group and individualized classes based on assessed needs and capabilities. Most major cities appear to offer small group training to explain use of their transit systems and answer questions. It appears to be important for acceptance by older people when transit systems stress some of the advantages of mass transit (greater safety compared to personal automobile travel, ability to avoid routine traffic, availability during disasters when roads can be blocked or clogged by traffic, increased fitness due to need to walk, etc.). It is also important to provide information about accessibility by people with limited mobility (e.g., elevators, kneeling buses, annunciators

Table 8.2 National Highway Traffic Safety Administration recommendations for model older-driver training programs

Successful programs will rely on effective informational and educational materials, using a variety of appropriate media, which:

- Facilitate self-regulation by sensitizing older drivers to the types of functional declines they may experience, and their consequences for safe driving.
- Provide advice and identifying resources to aid friends and family in problem identification and support for driving reduction/cessation.
- List and describe alternative transportation options specific to a community/county.
- Inform physicians of the driving risks associated with identified functional deficits, and describe feasible and standardized techniques for functional screening.
- Describe behavioral cues that police officers can use to identify at-risk older drivers, and procedures for referring suspect motorists for screening (in lieu of citation or other punitive actions).
- Provide easy-to-use tools for health-care and social services field personnel to identify gross impairments, guidelines for referrals for follow-on tests and/or remedial programs, and advice on issues of confidentiality and reporting to licensing authorities.

systems to announce stops). Finally, it is important to provide information about personal safety when using mass transit.

There is increasing recognition of the health benefit for older people of using mass transit, due to the need to walk to use it. A recent large study of over 7,000 people found that older people in England who did not use mass transit due to choice or to structural reasons (and not due to health-related reasons) had slower walking speed than older people who did use mass transit. The effect was seen on muscle strength in the lower limbs and not grip strength (upper body), suggesting an effect of walking per se (Rouxel et al., 2017). Physical fitness training may help offset these differences.

Public transport training should emphasize learning how to handle unexpected delays, canceled routes, adverse weather, and other contingencies. Mobile applications on a smartphone may help riders find alternative routes and determine time tables for delayed buses and trains. Older adults are also advised to have a contingency plan with emergency contact numbers (e.g., family, friends, or taxis) should they encounter unexpected difficulties during use of public transportation.

8.8 Route learning and navigation

Being able to learn new routes and navigate to specific locations (find one's way around) are essential skills for independent living. This is true regardless of the mode of transportation (e.g., driving, walking, or taking the train). These skills can decline in older adults for many of the reasons discussed in Chapters 3 and 4, such as deficits in working memory and in egocentric navigation. However, there is some promise that training programs may benefit older adults to both strengthen and maintain these skills.

For example, Mitolo and colleagues (2016) compared route learning training to an active control training that did not involve route learning for nursing home residents aged 70–90 years. The training was effective in improving navigation abilities among the residents immediately following training, and that improvement was still evident 3 months later.

Another approach to navigation training could involve teaching older people to use navigation software on smartphones. For example, Hickman et al. (2007) investigated how to best train older people on new technology so that they could later use the systems without instructions. They found that older adults particularly benefited from training that directed attention to a needed action rather than training that merely taught the needed action sequence.

In Chapter 4 we discussed the increasing use of smartphones by older drivers. The recent integration of smartphone navigation apps with vehicle console touchscreens (Apple CarPlay and Android Auto, discussed in Chapter 4) makes using smartphone navigation easier due to the projection of the maps on a larger display. A recent study found that use of navigation software with Apple CarPlay or Android Auto on the touchscreen caused less distraction than using a smartphone directly. However, when used to send texts or play music, that same system *increased* distraction during driving (Strayer et al., 2018). Thus, older people would benefit from the use of Apple CarPlay or Android Auto, but only to navigate. However, these systems are only available in cars built beginning in 2016.

8.9 Training for advanced in-vehicle technology use

The rapid proliferation of advanced in-vehicle technologies such as ADASs translates into the need for continual education or training on how to best take full advantage of these systems. Drivers of all ages will likely benefit from these new technologies, but the safety benefits for older adults in particular are profound. Specific technologies are discussed further in Chapter 11. Here we discuss recommendations for designing training in their effective use.

Little research currently exists identifying the most effective methods of providing older drivers (or drivers of any age) with education and training on how to use ADASs. One of the largest efforts to date to educate the general public about ADASs is a joint effort between the National Safety Council and the University of Iowa called the "My Car Does What?" project. (See Recommended Readings section for more information.) Additional efforts are underway, and current best practices from these are discussed here. However, further research is needed to ensure that older drivers are able to maximize their benefit from these new and evolving technologies. Since the systems vary from one vehicle to the next and even within one vehicle, updated training on these systems needs to be ongoing.

Most, if not all, vehicle manufacturers currently have online training in the form of videos supported by text content for how the major ADASs function. They vary in their coverage of the system limitations and the level of detail. Many people currently receive information about system use and function from individual sales representatives at dealerships; but, the detail and accuracy of this information vary widely. At present, though more research is needed, the best form of training appears to be a combination of one-on-one instruction from a knowledgeable person during a test drive or point-of-sale interaction supplemented by additional resource information for later individualized training.

AARP offers "Smart Driver Tech" online and in-person workshops that emphasize safety benefits of various technologies. See the Recommended Readings section for a website describing the training and for obtaining a list of locations for the in-person training. The training explains the capabilities and functions of a number of ADASs including blind-spot warnings, automatic emergency braking, forward collision warnings, and other components.

8.10 Determining fitness to drive

Several agencies publish guidelines to assist older adults and their family members with determining when it may be time for an older person to stop driving and to consider alternatives to the personal automobile. See the National Institute on Aging's (NIA, 2004) Older Driver report for a list of resources in this area. Table 8.3 lists questions the NIA report recommends that older adults ask themselves when considering whether it is time to stop driving. Rizzo (2011) provides an accessible review of advantages

Table 8.3 Questions older drivers should ask themselves

Is it time to quit driving?[a]

- Do other drivers often honk at me?
- Have I had some accidents, even "fender benders"?
- Do I get lost, even on roads I know?
- Do cars or people walking seem to appear out of nowhere?
- Have family, friends, or my doctor said they are worried about my driving?
- Am I driving less these days because I am not as sure about my driving as I used to be?

[a] The National Institute of Aging (NIA) suggests that if older adults answer yes to one or more of the above questions, they should seriously consider stopping driving. Find more information on their website: https://www.nia.nih.gov/health/older-drivers#give-up

and disadvantages of different methods of assessing the potential need for driving cessation and provides a list of medical conditions and their likelihood of resulting in the need to stop driving.

8.11 Design recommendations

Driving Training Programs

- In training programs, include some type of cognitive screening to ensure that trainees are functionally capable of benefiting from training. The Mini-Mental State Examination (MMSE) is a low-cost and easy-to-administer test that can screen for dementia or mild cognitive impairment (Folstein et al., 1975).
- Training programs should provide instruction on normal age-related changes that place older drivers at increased risk (e.g., visual and attentional changes, slowed processing speed, etc.).
- Provide instruction in traffic laws and other rules of the road including the meanings of signs and other traffic control devices, right of way, etc. This is important, as some traffic control devices have recently changed, notably devices for permissive left turns.
- Training programs should provide individualized feedback on performance to assist individuals to more accurately calibrate their perceived abilities with their actual abilities.
- At a minimum, older driver training programs should contain components including: (1) recognition of age-related changes in vision, cognition, and motor abilities that negatively impact driving; (2) speed of processing training; (3) training scan patterns at intersections; and (4) vision testing under low-light conditions.
- Older-driver training programs should include risk assessment. The most commonly used one to date has been the Driver Behavior Questionnaire (DBQ).

Transport Training

- Transport training programs should emphasize the advantages of mass transit, including such things as safety, increased fitness, and ecological benefits.
- Provide information regarding accessibility for people with limited mobility (e.g., access to elevators, kneeling buses, etc.).
- Provide navigation training and training on how to use mobile navigation resources on smartphones or computers.
- Ensure that older adults are provided with training on accessing assistance if they encounter difficulties (e.g., delays, canceled buses, or if they get lost or confused).

Recommended readings

AAA Senior Driving. Retrieved from http://seniordriving.aaa.com/

AARP Smart Drivertek. Retrieved from https://campaigns.aarp.org/ findaworkshop/ (provides training to members on new advanced technologies and ADASs).

An example of mass transit information on mobility. Retrieved from http:// rtachicago.org/rider-resources/accessible-transit/travel-training

An example of mass transit information on personal safety. Retrieved from https:// www.mta.maryland.gov/safety-quality-assurance-risk-management

Eby, D. W., Molnar, L. J., & Kartje, P. S. 2009. *Maintaining Safe Mobility in an Aging Society.* (In particular see Chapters 6–9, and Appendix A for information on Screening, Assessment, Licensing, Training, and Evaluation Tools for older drivers, respectively.)

Federal Motor Carrier Safety Administration (FMCSA)—for up-to-date information on how various medical conditions impact driving. Retrieved from https://www.fmcsa.dot.gov/regulations/medical/ reports-how-medical-conditions-impact-driving

Johnson, J., & Finn, K. 2017. *Designing User Interfaces for an Aging Population: Towards Universal Design.* Cambridge, MA: Morgan Kaufman/Elsevier. See Chapter 11, page 193, for an example of the design process for a public transport assistant for older adults.

Martinussen, L. M., Møller, M., & Prato, C. G. 2014. Assessing the relationship between the driver behavior questionnaire and the driver skill inventory: Revealing sub-groups of drivers. *Transportation Research Part F: Psychology and Behaviour,* 26, 82–91. doi: 10.1016/j.trf.2014.06.008

My Car Does What? A website designed to disseminate information about ADASs. For the general website, see https://mycardoeswhat.org/; for information specific to older drivers, see https://mycardoeswhat.org/ helping-older-drivers-stay-safe/

chapter nine

Alternative forms of transportation

Meet Bob and Lisa. Bob is 85 years old and recovering from a second stroke. He quit driving after his first stroke, 10 years ago, and has relied heavily on his wife, Lisa, who is 83 years old, as his partial caretaker and for all his transportation needs. Lisa maintains a valid driver's license, but she has increasingly been self-restricting her driving. She has never liked to drive in high-traffic areas, such as major urban areas and freeways. However, now she feels unsafe in these high-demand situations and further has tried to avoid driving at night, in bad weather, and for long distances.

A condition of Bob's release from the post-stroke rehabilitation facility he has been in for the last 3 weeks is a transportation plan to ensure that he is able to receive adequate follow-up treatment and other necessary health services. Aisha, a home-health nurse, is discussing the options available that Bob and Lisa may wish to consider. "First," said Aisha, "Most urban areas have a website providing transportation options. It will generally include descriptions and phone numbers you can contact for additional information. One of the first things you will want to do is make a plan for how you will get back and forth to medical appointments. For example, in Fairfax County, Virginia, the county maintains a website called, "Transportation for Older Adults and People with Disabilities" that can be found at the following URL: http://www.fairfaxcounty.gov/dfs/olderadultservices/transportation.htm. Here you will find options for free transport services for Medicaid recipients to Medicaid service providers, information about Travel Training to help you learn to use the public bus and metro systems, as well as information about local community programs that offer rides for older adults." Bob and Lisa, though fictional, are not unique. A growing number of people need alternatives to the personal automobile to meet their transportation needs. In this chapter, we discuss design issues for alternate forms of transportation, such as public buses and subways. We also discuss the design of community-based transportation services that may be staffed primarily by volunteers. We also cover accessibility and usability for web and mobile access to transportation resources. Finally, we briefly discuss applications of autonomous vehicles in the role of assistive technologies for older adults.

9.1 Designing public transportation to support older travelers

One option for many older adults who cannot drive, whether due to the loss of a driver's license, declining functional capacities, or self-restriction, but who want to maintain their mobility but lack the funds to hire a private driver are public transportation services. Though these services may be extremely limited in rural areas, they are common in urban areas. These services include buses, trains, subways, and tram systems.

9.1.1 Buses

In urban areas, buses are one of the more accessible systems for older adults, in that bus systems can be implemented anywhere where roads are already available so they do not need expensive or intrusive construction in contrast to trains or subway systems.

Often, due to the variability of traffic congestion at any given time, buses may be behind schedule. This means that passengers must arrive before the scheduled time of the bus's arrival, and may be waiting for a considerable amount of time. In the case of older adults, it is essential to provide some form of respite, such as a small bench. Preferably, bus stops should include a partially enclosed shelter to provide shade, and to shield travelers from wind or rain. As discussed in Chapter 7, older adults may be more wary than the general population at night, so all bus stops, whether they have a shelter or not, should also provide adequate lighting. Further design considerations include the provision of large, easily readable, and up-to-date route maps, timetables, and emergency contact numbers. Designers might also consider the addition of emergency help stations, or heaters if the route is in a particularly cold region.

Considerations for the design of bus systems themselves need to take into account the physical limitations older adults face. For example, older adults often have difficulty mounting high steps, such as are required to get onto many buses. To mitigate this issue, low-floor or kneeling buses have been developed. Low-floor buses are those where no step is required to board the bus, and this may apply to the whole bus or be low-entry only, where only the entryway of the bus is at floor height. Kneeling buses use hydraulic devices to allow the bus to "kneel" or lower down to street level to negate the need to step up to board the bus. Both of these types of buses would be conducive to easier boarding by older adults.

Older adults often also face problems with their balance, making it extremely difficult for them to traverse moving platforms, like a bus in transit. This means that drivers need to be aware of passenger placement and ensure that those with poor balance are properly seated before continuing with the route. When seating is unavailable, passengers need to hold onto rails or handholds to stay in place during transit. As discussed in Chapter 3, older adults often have range of motion limitations and may be unable to reach high into the air as is required for some handholds on buses. Therefore, it is important to ensure that lower handholds and/or adequate reserved seating (with appropriate signage) are available for older adults and persons with disabilities. Policies are needed to ensure that young healthy adults know such seats are reserved for older travelers. Adequate bus route maps should be posted and visible from anywhere on the bus. In the case of longer routes, it may not be feasible to make the font large enough to read from long distances; therefore, designers should consider ensuring that these are posted in multiple places throughout the bus. If possible, it is also recommended that digital readouts be available to list the current and next stops so that older adults do not have to rely on memory or hearing the conductor as stops are announced.

Route planners should consider the needs of older travelers when choosing an overall network design and for determining the stop density (i.e., the number and frequency of stops) and route frequency of the bus system. Generally, the higher the stop density, the more time it takes the bus to traverse a given route. At the same time, higher stop densities mean that individual travelers do not need to walk as far to reach a stop. Since older adults typically have difficulty walking long distances, higher stop densities will tend to favor older adults, providing greater ease of access and increasing the probability that older adults will view the bus as a viable option for their transportation needs. Route planners should consider increasing the stop density in areas with high numbers of older adults (i.e., near retirement communities and near medical facilities.)

If transfers between popular routes are required, timing of buses is a key issue. Route planners should ensure that there is adequate time for older travelers to egress one bus and walk to the location of the next bus stop. At the same time, long durations between buses and frequent changes will pose more challenge for older travelers and should be considered in route planning. Trip planning services should be provided and be accessible to older adults. These services should provide the total duration a trip is expected to take, how many changes may be necessary, and how long travelers have to make each change. These services assist all travelers but are particularly important for older adults who tire more easily and who may not be able to withstand long environmental exposure (e.g., heat and cold).

9.1.2 Subways and trams

Many of the issues relevant to bus design are similar to those pertinent to the design of subway and tram systems. Tram stops should consider the same design recommendations as bus stops: adequate seating and lighting, route timetables and maps, and shelter where possible. Because trams will be unlikely to have hydraulic systems to allow them to "kneel," steps with low enough railings for older adults to reach should be available. Sometimes, trams may also run in the center of a road, between directions of traffic. In this case, all tram stops should have crosswalks that help to ensure safe access from the sidewalks on either side of the road. See Chapter 7 for a discussion of design recommendations for crosswalks.

Subway stops generally differ from bus and tram stops in that they are almost always completely separated from normal traffic, are gated (meaning riders pay before boarding as opposed to trams and buses where riders either show prepaid cards or pay at the front of the tram or bus), often have a dedicated staff with informational facilities, and, in many cities, are underground.

Accessing the platform may be one of the more difficult issues that older adults face, particularly when platforms are underground or elevated, as entry requires the use of stairs, an escalator, or an elevator. In many older subway stations, elevators are not available at every entrance, or when they are, there is no clear marking to direct passengers. This is particularly apparent in New York City, where much of the subway system predates the Americans with Disabilities Act of 1990. Matthew Ahn, who holds the Guinness World Record for fastest time to travel to all New York City subway stations, noted that, as of 2015, little more than 100 of the 490 stations in the city were accessible to those with limited mobility (Ahn, 2015). He re-created the official map to show how it would look to someone who is wheelchair bound as compared with the official map. He further noted that there are five major subway and light rail systems in the country: Washington, DC; San Francisco, California; Chicago, Illinois; Boston, Massachusetts; and New York City, New York. Only Washington, DC, and San Francisco are fully accessible, while less than half of the stations in the Boston, Chicago, and New York areas are accessible. It should also be pointed out that while all stations in DC and San Francisco are designed to be accessible, elevators frequently break down and are not fixed for long periods of time, making regular use of these services by those with mobility limitations infeasible. In planning subway systems, designers should be certain to ensure not only that features to support accessibility are included, but that the features are redundant and have maintenance plans in place to ensure their continued usability.

A final matter to consider is how subway riders pay for their fares. Unlike other public transportation systems, subway fares are often collected by automated systems and in bulk, as opposed to individually by a fare collector. In these cases, where few if any staff members are available to assist passengers with obtaining a fare card or adding money to a fare card, it is of the utmost importance that fare systems be intuitive and easy to use. Designers should consider performing usability testing on fare machines to ensure that all riders can easily access services. At a minimum, to promote use by older adults, automated systems should provide clearly legible instructions and displays and provide menu selection and payment controls within easy reach to accommodate individuals with restricted range of motion.

9.2 Private transportation systems

Although public transportation is often the most affordable way for older adults to maintain mobility, it is unavailable in rural areas and many of the other major areas where older adults currently reside. This is due to the fact that many older adults choose to "age in place" or stay in their own houses rather than in nursing homes or assisted living facilities. Therefore, for a large portion of the population, older adults are staying in mainly suburban or rural areas where public transportation is either not implemented or not feasible. In these cases, often older adults will turn to private or community-based transportation systems, such as senior transportation services, friends and family, or even newer options such as app-based transportation services like Uber and Lyft.

9.2.1 Senior transportation services

Older people who do not drive still need to get to physician appointments, shop for groceries, etc. The federal Administration on Aging has the mission to carry out the Older Americans Act of 1965 by providing services designed to help older Americans live independently in their homes and their communities. One important use of Administration on Aging funds is to fund transportation services for people over age 60. The Administration on Aging administers a website, the "eldercare locator" (see Recommended Readings section) through which one can obtain phone numbers and websites of services in a given zip code. The ride services typically offered are shared rides, so they can take more time than private services. However, they are available in rural areas and are subsidized and therefore are also less costly than private transportation services.

9.2.2 App-based car services

Uber and Lyft are currently the two most common app-based car services. They are accessed mainly by smartphones (but can also be accessed from a computer-based website) and are intended to be used much like a taxi: the rider "hails" the Uber or Lyft driver by putting his or her address and destination into the app interface, and a driver who is in the vicinity picks the rider up. One advantage is that cash does not change hands as the rider's credit card is billed for the ride. The two downsides of these services for older drivers are that they require the use of a smartphone or access to a computer, and they are not intended to be used by, say, a family member who can hail the ride for his or her senior relative. However, companies such as GoGoGrandparent have been created to bridge this gap. GoGoGrandparent works with companies like Uber and Lyft, creating a bridge that allows users to call a ride from a landline, to request a ride for someone else, and to track someone else's ride to ensure their safe arrival. Both Uber and Lyft have also begun to attempt to integrate similar concierge and senior-friendly services. As mentioned in Chapter 3, Lyft has now partnered with Blue Cross Blue Shield Association (BCBSA) to provide transportation via Lyft for people insured with BCBS to the offices of health-care providers. Building on that relationship, BCBSA recently also partnered with Walgreens and CVS pharmacies to provide transportation via Lyft to those pharmacies in certain areas.

9.3 Designing effective web access to transportation resources

It is critically important that older adults be able to locate and access transportation resources. Often, older adults must rely on family members or communities to which they belong (such as religious groups or social organizations) either to provide transportation assistance or to recommend programs that may do so. Otherwise, older adults can request materials from organizations such as AARP or AAA by phone or mail. However, as we enter an age in which about 75% of the US population has access to the internet, increasingly more older adults will be consulting the web for the best transportation resources.

9.3.1 Effective or intuitive web access

One important aspect of effective and intuitive web access is accessibility. That refers to how well an intended user can navigate and gain information from a given website regardless of any impairments or disabilities the user may have. For all federal agencies, ensuring accessibility is a law covered under Section 508 of the Rehabilitation Act of 1973 (see https://www.section508.gov/ for more information), but most private corporations are under no such mandate to ensure accessibility. Luckily, there are many resources, and even entire companies dedicated to ensuring accessibility and usability for web content (see https://www.usability.gov/ for additional information).

As discussed in Chapter 3, older adults experience various sensory, cognitive, and physical changes that may impact their ability to use web or mobile resources. Specific limitations are presented in Table 9.1.

In summary, these functional limitations imply that older adults may have problems navigating websites that are otherwise considered to be highly usable by the general population. Where a website with animated

Table 9.1 Summary of age-related functional limitations

Function	Limitations
Vision	Decreases in near-vision focus
	Decreased color perception and sensitivity to color changes (where differences in the red/yellow end of the spectrum are easier to distinguish than blue/purple)
	Diminished contrast sensitivity and capacity to adjust to changes in light level
	Reduced visual field
	Increased likelihood of having cataracts or age-related macular degeneration (AMD)
Hearing	Hearing loss
	Decreased ability to hear speech in noise
Motor skills	Increased likelihood of tremors, rigidity, bradykinesia, and postural instability
	Increased risk of Parkinson's and arthritis
Cognitive abilities	Memory loss
	Confusion or problems following conversation flows or understanding speech
	Decreased ability to communicate effectively
	Increased distractibility
	Difficulty remembering names and locations

Source: Adapted from Arch, A. 2010. Web accessibility for older users: A literature review. Retrieved from https://www.w3.org/TR/2008/WD-wai-age-literature-20080514/

graphics, brightly colored text, and lots of content might be great for younger users, animations may confuse older users, colors used to emphasize text or clickable links may be indistinguishable from normal content text, and content that is too long or spread out across pages may be impossible to follow for some older adults. A great resource for ensuring accessibility for all users (not only older adults) is the World Wide Web Consortium (W3C: see Recommended Readings section).

When it comes to designing transportation-related resources for older adults online, the importance of accessibility for those resources is closely followed by their findability and the applicability of the content.

9.3.2 Search engine optimization

Search Engine Optimization (SEO) is a process by which the visibility of sites online can be manipulated. When sites have not been optimized, they may be difficult to find, particularly for users who do not know what search terms are needed. Often, when any user begins to research a topic, he or she will open a search engine and input the terms that he or she thinks are necessary. For example, in the cases of Bob and Lisa, Lisa may input the terms "driver assistance in Fairfax in the evening," but the most relevant site for her might not appear because it has not included any of those keywords in its content. Lisa may spend hours and find some sites that are similar, but unless she changes her keyword search to the right combination of terms, she might never see certain sites that could match her needs. Google offers an SEO starter guide that can help with SEO for developers, or various companies offer SEO services such as tools, training, and technical advice (link provided in Recommended Readings section).

9.3.3 Transportation resource needs

The final component of a useful transportation resource site for older adults is the content. Designers could use an initial page with an uncluttered layout that either directs users based on their needs or asks a series of questions to determine which content to display to the user.

9.4 Future autonomous technologies

In designing for the future, planners may wish to consider planning for the incorporation of technologies that are currently in development. Notably, autonomous vehicles have the potential to increase mobility for older adults. Imagine, for example, that Bob and Lisa do not have to rely on Lisa to drive Bob to his therapy appointments, but that Lisa could summon an autonomous vehicle to pick Bob up. Or that Bob and Lisa could purchase a new vehicle that could detect Lisa's discomfort and even take over control of the vehicle when she feels unable to handle driving conditions. Furthermore, autonomous technologies could allow completely unmanned subways, buses, and trams, increasing the availability and hours of these services with modest increase in costs. As of this writing, Waymo is offering a commercial autonomous ride-share service in the Phoenix area. Regardless of the success of this and similar efforts, it will continue to be imperative that designers create transportation systems that cater to users of all ages, including those with disabilities.

9.5 Design recommendations

- Buses
    - Ensure bus stops have seating and lighting, preferably with some coverage from weather.
    - Provide large, readable, intuitive maps, timetables, and emergency contact numbers.
    - Equip buses with low steps or hydraulic kneeling systems, and ensure railings and handholds are available, in compliance with the ADA.

- Provide adequate seating with clear signage for older adults and persons with disabilities.
- Provide on-board digital displays of the vehicle's current location and the next stop.
- Subways and trams
  - Provide crosswalks and pedestrian refuge islands for trams that stop in the median between traffic lanes.
  - Provide access to well-maintained elevators at all under- or above-ground stations.
  - Provide intuitive payment systems and information desks where possible.
- Web access to transportation resources
  - Adhere to established accessibility guidelines including Section 508 compliance.
  - Consider SEO when designing sites.

Recommended readings

Eldercare locator. https://eldercare.acl.gov/Public/Index.aspx

Federal Highway Administration (updated September 5, 2017). Transportation Alternatives. Retrieved from https://www.fhwa.dot.gov/environment/transportation_alternatives/overview/

World Wide Web Consortium. 2008. Web content accessibility guidelines (WCAG) 2.0. Retrieved from https://static.googleusercontent.com/media/www.google.com/en//webmasters/docs/search-engine-optimization-starter-guide.pdf

chapter ten

Designing aviation travel services for older passengers

Imagine a person named Mariam. Mariam is 32 years old and is traveling to visit an aunt who lives across the country. Mariam knows what day she needs to leave and what day she wants to come home and uses a web-based travel site to find a ticket for an aisle seat on a nonstop flight. On the day of her trip, Mariam leaves her house and uses her GPS system on her phone to navigate to the airport. It takes her an hour to get there, with no navigational hiccups. However, once on airport property, the GPS no longer recognizes the maze of winding roads, and Mariam must follow airport signs to find the long-term parking lot. The first sign set differentiates the five lanes of traffic into three lanes for Arrivals and three lanes for Departures with one dual lane splitting off in both directions. The second set of signs lists parking lots—hourly, daily, short term, long term, and economy. Mariam finds her way to the long-term lot and unloads her luggage. She must next find and board the shuttle to the terminal. She then walks the 200 yards to the departures entrance and is confronted by a line 25 people long waiting to check their bags. Luckily, her airline at this airport has self-help kiosks. Mariam steps up to a kiosk with a computer and is asked to identify herself to the system by entering her nine-digit confirmation number, her passport number, her ticket number, or to look up her information using the four-digit number of her flight. She pulls out her phone and looks up her confirmation number and enters that into the system. She then answers a series of questions related to the shape, weight, and content of her luggage as well as her seating and printing preferences for her ticket. After completing this task, she receives her boarding pass, her luggage receipt, and her luggage tag. She then continues to the counter where she has only five people to wait behind.

Once she has checked her bag, she must go through security. She follows the signs for screening and waits in the line to have her ticket and identification checked. This takes approximately 10 minutes. The line for screening then takes another 25 minutes, moving slowly as passengers take off their shoes and jackets, and place their personal items and carry-on luggage on the belt. Once she has cleared security, she follows signs for the shuttle to her terminal and waits the 3 minutes until the shuttle arrives. In the terminal, she finds the departures board and confirms her gate and flight time. Her gate is 17 gates away from the shuttle drop-off point, and her flight departs in a little over an hour. She stops to get a drink and snack before continuing to her gate. It takes her about 15 minutes to walk there, and by the time she has arrived she has half an hour until boarding. As all of the chairs in the small waiting area are full, she sits down on the floor near a pillar and reads her book. She becomes so engrossed she almost does not hear the final announcement that her group is boarding. She boards the plane, finds her seat, and hoists her luggage into the overhead storage area. She finally sits down in her small seat in coach, relaxes, and mentally prepares for the 6-hour flight. Now, imagine that it is not Mariam visiting her aunt, but the other way around. Her aunt Kate is 73 years old. She is retired after 43 years of teaching fifth grade. She has some hearing loss, and her left knee is still a little weak after a partial knee replacement 8 months ago. Imagine the challenges Kate might experience with the aforementioned demands of air travel, including flight booking, driving to the airport parking lot or departure curb, checking luggage, clearing security, finding the gate, boarding the plane, and enduring a 6-hour flight.

Older adults in developed countries are healthier, more educated, and wealthier than any previous generation. This has led to an upsurge in the amount and duration of air travel trips undertaken by older adults. The term *third age* has been coined to describe this postretirement period in life when active older adults spend their time traveling to pursue self-development and leisure activities. However, aviation services have been slow to catch up with the needs of aging travelers. In the case described, the process may be very mentally and physically demanding for older travelers. One aid is the increasing use of smartphones and tablets by older people (see the following text), which opens to them the advantages of computerized travel services.

10.1 Booking a flight

Booking a flight can be done by phone or via websites. Internet usage has grown sharply among older people in recent years reaching 66% of those aged 65 and older in 2018 according to the Pew Research Center. An older person who is not computer literate can still make flight reservations by telephone, though there are often fees associated with doing so. Speaking to a human agent will ensure the reservation is made correctly and will allow the older passenger to request any needed services such as wheelchair use and early boarding. Older people who are computer literate can book their flights on airline websites. As of 2016, the Air Carrier Access Act required airline websites and computers to meet the accessibility standards of the Website Content Accessibility Guidelines (WCAG) (link provided in Recommended Readings at the end of the chapter). Most airlines allow cancellation of a booking within 24 hours without a penalty. There are also universal airline booking sites through which many airlines can be booked. Those can have rigid cancellation policies.

Smartphone- and tablet-based travel websites and applications can benefit older travelers. According to the Pew Research Center, about half of those older adults who do use cellphones have a smartphone, a figure that has doubled since 2013. Smartphones are owned by 59% of 65- to 69-year-olds, though that number drops to 49% among 70- to 74-year-olds. Tablets are owned by about 33% of older people, but that rises to 62% in older people with incomes of $75,000 or more. Travel websites and smartphone applications allow users to input various aspects of their personal information (e.g., storing a "Home" location, flight information, or bus route information). After information has been stored, the application integrates important aspects of travel. These applications can track flights with up-to-date information about gates, boarding times, landing times, and delays; tell the users when they should leave for the airport based on their location; and offer navigation routing (or link them to a taxi service). Once the user has reached his or her destination, the application can present map and navigation information and give recommendations for restaurants or attractions. Additionally, e-tickets and passes can be stored electronically, eliminating the need to print boarding passes. This reduces the memory load for the passenger insofar as gate information can be stored and updated on a smartphone or tablet. Moreover, gate changes and flight delays can be easily checked by passengers on their smartphone or tablet.

It is important to consider the particular needs of older users when designing travel websites for computers, smartphones, and tablets. Smartphones have small screens on which maps and text may be hard to read in bright sunlight, though font size is easily adjusted. Typing on the relatively small keyboards of smartphones can be challenging for people with poor dexterity. Alternatively, travel applications can also be accessed via tablets, which have larger screens and might be a better choice for an older traveler. However, as tablets cannot fully replace a phone, the traveler would need both expensive devices.

Some older travelers may need to request additional services for airline travel. Some older adults may require a mobility aid (defined by the Americans with Disabilities Act to include canes, crutches, or wheelchairs) or an electric cart to move through the airport. Those travelers may have impaired cognition, impaired hearing, or low vision. They may need oxygen tanks or special medications that need to be accessible during the flight. People who have had knee or hip replacements need to be aware that they should inform the security agents in advance of that fact, as they must go through a specific metal detector to avoid a long delay if their device sets off an alarm. Older passengers may have slight cognitive deficits rendering an airport environment (almost always characterized by bustle, noise, and countless signs) stressful and confusing.

Airline websites do allow the traveler to request specific services such as mobility aids or early boarding. However, there is little standardization for design of airline websites (beyond the accessibility requirements), and the appropriate links to assistance options may not be easy to find. Passengers can also notify the airlines via the website if they have peanut-dust allergies, if they will be traveling with oxygen tanks, or a "trained assistance" animal, etc. Passengers can also create an account with a given airline and set up a "profile" that stores preferences and needs for specific assistance. Displaying a list of options that may be frequently used by older travelers after a traveler types in date of birth could facilitate use of such additional services (early boarding, extra leg room, mobility aid, etc.). It would also be useful for the travel website to provide maps of airports, showing parking options, and with estimated walking distances between ticketing services, security checkpoints, and gates. Such information can assist older travelers with determining whether they will need to request a wheelchair or a cart when they book their flight.

10.2 Getting to and into the airport

Although many aspects of air travel are highly regulated and have little variation between airports, the layout of an airport, both inside and out, can vary dramatically. However, almost every airport can be accessed by personal vehicle, bus, taxi, and, in some cities, light rail or subway systems. Incorporation of design recommendations for older adults using each of these travel modes can greatly improve their travel experience.

10.2.1 By personal vehicle

Driving oneself to the airport can be a challenge for individuals of all ages. Do you want valet, short-term, long-term, or extended-term parking? Various terms are used for these options, for example, daily, hourly, and weekly. What are the fees associated with each parking option? Do you pay when you enter or when you leave? Is there valet parking, and what is the cost? Is there a bus or a tram from a given parking area to the terminal, or will you be required to walk (with all your baggage)? Just the act of reading and following signs to reach the correct lot can be a daunting task. Figure 10.1 presents an example of the typical level of complexity found at many airport entrances.

Airport designers must consider older drivers as well as the general population when designing signage. As discussed in Chapter 3, older adults have decreased contrast sensitivity, take longer to extract visual information from signs, and have more difficulty searching for a particular item in a cluttered background. Airport roads are similar to work zones

Figure 10.1 The Tom Bradley International terminal in Los Angeles, California, which has the most origin and destination flights in the world. (This file is licensed under the Creative Commons Attribution-Share Alike 3.0 Unported license.)

in that the information provided can be novel. Work zone signage was discussed in Chapter 5. In summary, message signs should be made of retroreflective material, be elevated over the roadway, and have lighter-colored letters on a darker background. Additionally, important information (e.g., airlines, terminals, and parking) should be redundant (repeated at different locations) in case some of the information in the first sign is missed or unclear. Reasonable guidelines, following those used for the design of international airports, are given in Table 10.1.

Table 10.1 Recommended text height for airport signage

Area	Recommended text height
Wall plaques including restroom, stair, and elevator identification	5/8″ text height
General pedestrian overhead signs	3″–5″ text height
General vehicular signs in low-speed areas	5″–8″ text height
General vehicular signs in high-speed areas	10″–12″ text height

Source: Adapted from Washington Dulles International Airport Design Standards and Signing Guidelines.

10.2.2 Airport parking

Parking lots should have signage indicating the locations of elevators, shuttle stops, directions to access the terminal, and emergency exits. Reminders to take one's parking entry ticket as well as to note down or photograph parking locations could help older travelers to locate their vehicles in the parking garage after they return.

When parking lots are far from the airport terminal, buses or shuttle services are usually provided. Shuttle stops should be easily identified, and information should be provided regarding the relative frequency of service. Shuttle stops should have some benches, be protected from weather, and display emergency safety telephone numbers. Shuttle stops in long-term parking lots should be located to ensure that older travelers do not need to walk long distances from their car to reach the shuttle stop. In all lots there should also be phone numbers displayed so that travelers can request additional information or accessibility options (such as shuttles, wheelchairs, or other mobility aid).

10.2.3 By taxi or friend/family member

Many older adults access airports by being driven by friends or family members or car services such as taxis and ride-hailing services. This spares the older people the difficulty of transporting luggage from a parking lot, even with adequate shuttles. Airport designers must consider older adults' physical capabilities when designing drop-off points. There should be adequate space for vehicles to pull completely clear of the roadway when offloading passengers. Airport services often (not always) include luggage porters who can remove luggage from vehicles and escort passengers to the check-in counter. Some airlines provide "skycaps" at the arrivals curb who require a small fee to help people check their luggage and obtain boarding passes.

10.2.4 By rail or bus

Public transportation is one of the few inexpensive forms of transportation to airports. Yet, adults over age 60 are less likely than younger people to use public transportation due to a combination of income and health. Those older people who are wealthier as well those in poorer physical condition are less likely to use mass transit (Zwald et al., 2014). People over age 65 are also more likely to live in rural areas, where public transit is less available. A number of US cities now have affordable light-rail systems that service airports. Older adults using light-rail or bus to airports would not need help with luggage until they reached the terminal. That may not be a problem with rail as most luggage has wheels, but it could be a problem with buses where passengers may need to lift bags into the bus and possibly into overhead storage areas.

10.3 Moving through the airport

Airport designers have at their disposal various human factors tools with which to analyze the flow process of passengers from curb to boarding. One resource is the recommended reading *Airport Cooperative Research Program (ACRP) Synthesis 51* report by Mein et al. (2014). Such analyses have been used to evaluate the efficiency of operations including check-in, security screening, and boarding. Similar analyses could be used to determine the flow process specifically for older passengers and those with disabilities to determine where confusions, bottlenecks, and delays might occur for such passengers.

Once a passenger enters an airport, accommodations for persons with disabilities are governed in part by the Air Carrier Access Act. The rules are aimed at preventing discrimination in air travel on the basis of disability (details available at the Department of Transportation [DOT] website: https://www.transportation.gov/airconsumer/passengers-disabilities). Older people who require additional services (mobility aid, early boarding, or extra legroom) should request them when booking the flight but can also request them at the airport. Many airports have skycaps at the arrivals curb to help with the luggage check-in process and porters and carts to get passengers to the gate. A number of airlines offer a fee-for-service scheme whereby the passenger is met at the curb and escorted through security to the gate (e.g., American Airlines has "Five Star Service"). In addition, there are companies that are not affiliated with a specific airline offering similar services. Such services can smooth the path of an older passenger through the airport.

10.3.1 Checking in

When flying on an airline that provides skycaps, luggage can be checked and a boarding pass obtained at the curb. Passengers and their luggage can also be checked in at the airline ticket counter without the need for computer literacy on the part of the passenger. However, one needs to stand in line for skycap and counter service. A faster process involves electronic check-in systems, whereby passengers can initiate the check-in process at home (on a smartphone or computer) but check their luggage and obtain a boarding pass at a kiosk in the terminal (Figure 10.2). An agent may still be involved, but the process is faster because the passenger checks in and obtains a baggage tag and boarding pass before seeing the agent, As previously mentioned, DOT rules require that airline websites and kiosks be accessible to passengers with disabilities and must meet the Web Content Accessibility Guidelines (WCAG) 2.0 Level AA standard (published as a W3C Recommendation, discussed in Chapter 9). Yet, the

Figure 10.2 Self check-in at Dublin Airport, Ireland. (This file is licensed under the Creative Commons Attribution 2.5 Generic license.)

use of kiosks can be challenging for passengers who are not computer literate. Agents are usually present near kiosks in self-check and self-tag areas, and such an agent could be important to help older people with low vision and with dexterity problems. Having a smartphone or tablet can facilitate this process as the boarding pass can be obtained before leaving home and be displayed on the device.

10.3.2 Security

Until recently, security procedures were cumbersome and uncomfortable for older adults. However, in 2012, the Transportation Security Administration (TSA) implemented new guidelines with specific restrictions and rules being modified for older adults (75 years of age and older, identified visually) who are now allowed to leave on a light jacket and shoes during screening.

Specific changes included allowing older adults to keep shoes and light jackets on during screening, and allowing older adults a second pass through imaging technology to resolve issues, instead of immediately continuing on to a time-consuming and intrusive pat-down. People are allowed to go through security in a wheelchair but then are subjected to a pat-down, and any removable pouches are x-rayed. A person who cannot stand in a security line for any period of time would be well-advised to request a wheelchair in advance from the airline.

10.3.3 Terminal amenities

Moving through the terminal can be challenging for older adults who nearly always have somewhat poorer balance than younger people. Older people who use a mobility aid may find escalators and moving walkways difficult to use. Elevators should be collocated with stairs and escalators and both should be indicated by signage.

10.4 At the gate

A typical airport gate waiting area has seating, a help desk, and some manner of information display relating to the status of the flight and plane to which the gate has been assigned. Design of the gate area should ensure that seating is spaced adequately to allow for easy movements for those passengers who use mobility aids and that there is ample seating for all passengers.

A problem for passengers of all ages concerns the almost ubiquitous use of spoken announcements of boarding and changes or updates to flight status. As spoken announcements are made at each gate, when announcements from adjacent gates overlap in time, it can be difficult to hear clearly. Most airlines use visual message boards near the gate, and those should be updated in concert with the spoken flight announcements.

Each airline has its own smartphone app that shows specifics on the flight, including the gate and any delays. For passengers with smartphones or tablets, such an app can be very informative, as it can be consulted even in the air during a flight. Some of these apps display maps of the airport, which can be very helpful in planning how to get from one gate to another.

10.5 Health and safety on-board

The problems discussed regarding the needs of older passengers may be compounded in the small space of the flight. Older adults often have medication that must be taken at a particular time, may need to use the restroom more frequently, or may need to move about more often to reduce the risk of blood clots. Additionally, as discussed previously, older adults are more likely to need mobility aids and/or oxygen tanks, which may make navigating a narrow plane aisle difficult. Such passengers can request early boarding when they book their flight. During early boarding the gate agents will help with boarding, including help with wheelchairs and with placing luggage in overhead bins. Another advantage with early boarding is that passengers with low vision and limited mobility can be seated near the front and more easily request help from flight attendants during boarding and during the flight. Passengers with low vision might have difficulty discriminating the icons for the flight attendant call buttons from the reading light. Passengers with flexibility limitations might have difficulty reaching the air nozzles and the flight attendant call buttons.

10.5.1 Connecting flights

Connecting flights are challenging for all passengers. Airports and airlines could use process flow analysis and task analysis to improve the experience of connecting flights for older passengers and those with disabilities. Procedures for connecting flights range from easy, where arrival and departure gates are near each other and there is plenty of time, to difficult and stressful, where arrival and departure gates in large airports may be more than a mile apart with very little time between flights. When a flight is delayed, flight attendants sometimes announce the gates of connecting flights prior to landing and make that information available after landing. However, usually the burden is on the passengers to determine their current gate, and their destination gate (which may have changed during the time that they were in the air). The passengers must then determine the route between gates and follow that route in what may be an unfamiliar airport. Under those conditions, older passengers may be worried about missing a connection. To make matters worse, when flights are international with national connections, passengers must often collect their luggage, re-check it, and then negotiate security a second time. While most airlines have assistants to help passengers with this, their services need to be requested in advance. Older adults or passengers with disabilities should request these services when they book their flights.

10.6 Leaving the airport

After landing and disembarking at their destination, passengers must follow the signs to baggage claim and to the exit for their ground transportation. Ideally, baggage is delivered in a timely manner to a baggage claim carousel which is well marked with flight numbers. Once passengers have found the correct carousel and baggage begins to be delivered, passengers must visually scan bags as they pass by at moderate speed to identify their own baggage. Once baggage has been spotted, it must be quickly lifted off the moving belt, a maneuver that may be difficult for older adults with large baggage. The older person would be advised to find a luggage porter at this point, as the porter can lift the luggage off the carousel and load it onto a wheeled cart for a small fee. Then the porter will move the luggage to the curb for the ground transportation. Alternatively, most airports have luggage carts that can be rented for a modest fee to get the luggage to the curb.

Baggage claim carousels should be as low to the ground as possible to ensure that all passengers can easily lift off their luggage. Often there is relatively little seating for passengers in baggage claim areas, which could be a problem for older people who have trouble standing for long periods. Once baggage has been claimed, passengers must obtain ground transportation to their final destination. Most airports have signage to direct people to the location of transportation services such as car rental desks, taxi stands, and parking lot shuttles. Again, use of a luggage porter (who must be tipped) can be helpful in getting the passenger's luggage to the correct service.

10.7 Conclusions

Air travel is a chain of events that must occur in a specific order, but one that has many opportunities for difficulty for the older traveler with or without disabilities. The Air Carrier Access Act (fully implemented in 2016) has resulted in improved access of airline and airport services for people with disabilities, including those that are common in healthy older people. In this chapter we summarized the particular problems faced by older passengers at each step in the process, from booking, to check in, to boarding, during flight, during flight transfer, and baggage claim. Many of the known problems arise due to age-related limitations in vision, manual dexterity, limb mobility, and balance. We pointed out services offered by airlines that can smooth the path of the older traveler, from human agents (phone or in person), skycaps, porters, websites, kiosks that meet accessibility standards, to wheelchairs. We also pointed out steps that the older traveler can take to smooth his or her own path during booking, such as requesting early boarding and extra legroom; when arriving at the airport, such as using skycaps and valet services; and at the terminal, such as using smartphones and tablet computers that facilitate boarding passes and gate changes. Many of the suggestions made in the chapter are implemented in most airports, though some are not implemented widely.

10.8 Design recommendations

- Booking
  - Travel websites and applications should incorporate general usability principles into the design of electronic booking systems, including ease of navigation, clear help features, visible progress status, and simple error prevention and detection. These should meet the Web Content Accessibility Guidelines (WCAG) 2.0 Level AA standard (published as a W3C Recommendation, discussed in Chapter 9).
  - Airlines could waive the standard phone booking fee for passengers with disabilities for whom booking online would be difficult.
  - Airline websites could display a list of options, based on passenger age, that may be needed by older travelers during booking (early boarding, extra legroom, mobility aid, etc.).

- Airline websites could improve the salience of links through which special services are requested and inform people requesting wheelchairs that they can also request an escort pass for an able person to accompany them to the gate.
- Older travelers who are computer literate could use a smartphone or tablet while traveling so they have easy access to gate information and boarding passes.
- Arriving at the airport
  - Airports could analyze the flow process of older passengers (including those with disabilities) from curb to boarding to look for ways to improve their experience.
  - Airlines could make it easier for passengers to determine whether skycaps are available at a given airport, as porters and skycaps at the curb can help older people and those with disabilities check their luggage.
  - Airlines could inform older passengers about services by which they will escort a passenger through the airport for a fee.
- At the gate
  - Airlines could provide seating labeled for the use of older people and people with disabilities near the visual message boards to allow them to more easily obtain updated flight information and gate changes.
- On the plane
  - Older people who have mobility problems or who have disabilities should request early boarding to obtain seats with extra legroom and to obtain help from flight attendants during boarding, during the flight, and during disembarkation. Most airlines allow passengers to request an escort for connecting flights.
  - Most airlines have free apps designed for smartphones that can show up-to-the-minute information on flight delays, gate changes, etc. These apps could be modified to add an estimate of the walking distance between gates and show the locations of moving walkways between two given gates. Most airline apps already provide maps of airports.
- Leaving the airport
  - Older passengers with disabilities should consider arranging a luggage porter to assist them with getting luggage to the curb.
  - Most baggage claim areas have only a small number of benches. Some of those could be designated for older people and/or those with disabilities.
  - Baggage claim carousels do not always have up-to-date information concerning the flights served at a given moment.
  - Lowering the height of baggage claim carousels would benefit older people less able to lift heavy baggage.

Recommended readings

Abeyratne, R. I. R. 1995. Proposals and guidelines for the carriage of elderly and disabled persons by air. *Journal of Travel Research*, 33(3), 52–59.

Passengers with Disabilities. Retrieved from https://www.transportation.gov/airconsumer/passengers-disabilities

Sproule, W. J. 2011. Airport planning for older air passengers. In *Reston, VA: ASCE Proceedings of the First Transportation and Development Institute Congress*; March 13–16, 2011, Chicago, Illinois.

Web Content Accessibility Guidelines. Retrieved from https://www.w3.org/WAI/standards-guidelines/wcag/

chapter eleven

Automated vehicle technologies and older adults

In many places throughout this book we have discussed the potential for new in-vehicle automated systems (often referred to as advanced driver assistance systems [ADASs]) to improve the safety and mobility of older adults. In this chapter we focus solely on the design of new and emerging in-vehicle technologies aimed at meeting the needs of the aging driver/ passenger/operator. The following fictional persona story illustrates how some of these technologies may change transportation.

Eiza is 84 years old and lives by herself in an urban townhouse. After several close calls and then an automobile crash a few years ago in which she received only minor injuries but totaled her car, she became convinced that it was time for her to give up driving. At first she was very concerned that this would mean the end of her active social life—meeting friends for dinner, the theatre, concerts, ballet, etc. But, then a friend introduced her to an electronic AutoCar application where she could use her phone to schedule a driverless vehicle right from her phone 24/7. Nervous about it at first, she has been using it successfully for a couple of years now. Her friends and family are no longer concerned about her safety while driving, and all are grateful for the mobility she maintains.

Eiza's phone rings, and she takes the call before the second ring, seeing from the caller ID that it is her friend Barbara. "Hey, Barbara. What's up?," Eiza answers.

"Hello, Eiza. How is the day treating you? Look, I was wondering if you'd like to meet for a drink at the Beach Shack next to Argio's before meeting up with everyone else for dinner so we have a chance to catch up?" Barbara asks.

"Sure," says Eiza, "that sounds great."

"Super," says Barbara. "Dinner is at 8 p.m., so let's meet about 6:30."

"Great, see you then." Eiza hangs up the phone and switches to the calendar application (app) on her phone, which is synced with her AutoCar app. She enters, "Beach Shack, 6:30 p.m." in her calendar and a message pops up asking if she'd like to update the AutoCar service. She selects "yes," and a confirmation pops up from her AutoCar service reading, "AutoCar to arrive at 6:20 p.m. en route to Beach Shack, 729 Washington

Blvd?" Eiza confirms the new pickup time and location and returns her attention to the manuscript she'd been reading on her laptop.

At 6:10 p.m. that evening Eiza receives a reminder notice on her phone from the AutoCar app that the car will be arriving in approximately 10 minutes. She hits confirm, puts the phone down, and checks herself in the mirror. Good to go, she thinks. A few minutes later another "Bling" "Bling" on her phone notifies her that her scheduled car is pulling up outside. She double-checks to make sure the cat is inside, locks the back door, and goes outside to meet the AutoCar. A pleasant electronic male voice with a slight British accent greets her as she approaches the door. "Good evening, Eiza! Out for a bit of fun, are we?" The rear passenger side car door opens automatically and Eiza seats herself inside. The passenger seat adjusts automatically to Eiza's prespecified preferences, and her most recent playlist starts playing on the car's audio system. "May I escort you to the Beach Shack now, Eiza?" asks the same male voice. Eiza taps "confirm" on her phone and the doors lock and the vehicle starts and begins navigating to the destination. The first few times Eiza took the driverless car—it does not have /an operator or a steering wheel—she was a bit nervous. Now she thinks to herself, "this has really simplified my life, and it is so nice to be able to meet up with friends without having to ask for rides." After arriving at the destination, the seat belt rescinds, the doors unlock, and at the same time that the message appears on her phone the electronic voice says, "Have a wonderful time, Eiza. May we pick you up at a specified time or would you like to send word when you are ready?" Eiza selects, "Schedule later," and gets out of the car to see her friend Barbara walking up the sidewalk.

Clearly the AutoCar application does not exist yet. And yet, it is conceivable that it could exist in the near future. Recently, Waymo began operating a paid driverless taxi service using autonomous minivans in Phoenix, Arizona. This functions much like the existing operator-driven ride-share programs Uber and Lyft, but operates only in one geographic area at present. How can such future systems, and also more near-term systems, be designed to meet the needs of older adults?

Before beginning our discussion, some terms are worthy of clarification. The National Highway Traffic Safety Administration (NHTSA) and SAE International have developed taxonomies for varying levels of automation. As shown in Table 11.1, Level-0 is used to designate fully manual driving with the level of automation increasing to Level-5 which designates fully autonomous driving. Self-driving vehicles would likely address most, if not all, the mobility challenges of older adults who could afford them. However, this level of automation is currently a number of years away for most people. Currently, and for the foreseeable future, we are dealing with intermediate levels, often referred to as ADASs. An ADAS requires driver monitoring and engagement by the driver, who retains the responsibility

Table 11.1 Levels of automation established by SAE International

Level	Label	Description
0	No automation	The driver performs all driving tasks
1	Driver assistance	Some driving assist features may be included
2	Partial automation	Vehicle has some automated features (e.g., acceleration and steering), but driver must always remain engaged and monitor environment
3	Conditional automation	Vehicle is automated, but driver must always be ready to take control of vehicles with notice
4	High automation	Vehicle can perform all driving functions under certain conditions
5	Full automation	Vehicle can perform all driving functions under all conditions

Source: Adapted from NHTSA. Automated Vehicles for Safety. https://www.nhtsa.gov/technology-innovation/automated-vehicles-safety

of taking over control when the ADAS fails or encounters situations outside its limits. As we pointed out in previous chapters, older adults have the potential to benefit greatly from ADASs. However, in order for them to have benefit, ADASs must be valued and accepted by older adults. Additionally, it is essential that instructions and displays be designed to support development of an adequate mental model of system capabilities and limitations. Recent surveys (McDonald et al. 2018) indicate that older drivers lag behind their younger counterparts in their expressed willingness to use advanced vehicle automation.

Difficulties that older drivers face on the road were discussed in Chapter 6. Recently, Bellet and colleagues (2018) used a focus group of older drivers to determine their perceptions of the driving situations they found most challenging and, subsequently, their perceptions of the usefulness of an ADAS that provided support for those particular aspects of driving. In general, older drivers expressed preference for systems that provided information to supplement their manual driving abilities rather than features that automated the task. For example, the older drivers expressed a high degree of perceived usefulness for a system that would tell them the current speed limit or alert them when they were exceeding the legal limit, but they were somewhat less favorable toward using automated systems like cruise control or adaptive cruise control. Interestingly, despite the risk posed by intersections and specifically making left turns (based on crash reports), older drivers did not perceive these situations to be particularly challenging. This mismatch between their perceived and actual abilities may be one of the reasons why older adults have a higher proportion of left-turn crashes relative to their younger counterparts. Despite the fact that older drivers in

Bellet et al.'s study did not perceive these left turns as particularly challenging, most were receptive to an automated aid to support them in these maneuvers. This was particularly the case if the system contained a warning for times when they had chosen an unsafe turn gap. Clearly in order for ADASs to have benefit, they must be used. Further research is needed regarding the best way to design training materials for how to use these systems and education to assist older drivers in understanding the potential benefit these systems can provide. Since new technologies continue to be developed and evolve, even postdeployment training will need to be an ongoing process, a topic that was discussed in Chapter 8. We continue our discussion of the design of these systems by examining common individual ADAS components.

11.1 Collision mitigation including emergency braking

Collision mitigation systems are designed to reduce the likelihood or severity of forward collisions. Different automobile manufacturers use different specific technologies, but in general they work through a combination of sensors (e.g., radar or lidar sensors located on the front or rear of the vehicle) and automatic and/or supplemental braking. Some systems also tighten the seat belts when risk of a crash is increased. As discussed in Chapter 4, older drivers have higher crash rates per mile driven than middle-aged drivers and due to their greater frailty are more likely to be seriously injured or killed when they are involved in a crash. So, any reduction in crash occurrence and severity has positive ramifications for older drivers and older passengers. Further, evidence indicates that collision warnings can benefit older drivers even more than young drivers—acting as an extension of their perceptual capabilities to alert them to events they might otherwise miss. Presenting precollision warnings in multiple modalities can offset age-related slowing in processing speed, allowing older drivers time to process the warnings.

Automatic emergency braking systems apply the brakes only if the driver fails to do so. Such systems do not prevent the driver from braking. The driver feels the brake pedal being depressed by the automation. Such systems are not guaranteed to prevent a crash but rather designed to mitigate a collision either by preventing the collision or by slowing the speed of the collision. The important advantage of automatic emergency braking over collision warning systems is that the driver does not need to attend, or make a response, or even be conscious for emergency braking to occur. However, typically the collision warning occurs prior to the automatic braking. Another advantage is that automatic emergency braking systems are difficult to turn off by mistake. The controls for automatic emergency braking can typically only be accessed through menus (not controls on the dashboard).

11.2 Blind-spot monitoring

Lane-changes are a dangerous maneuver during driving (Chapter 5). Checking for vehicles in one's blind spot before changing lanes requires a trunk and head rotation that is difficult for older adults due to their restricted range of motion, as discussed in Chapter 3. This is likely a contributing factor to why many older drivers use their turn signals but neglect to check their blind spots before making a lane change maneuver. Failure to check blind spots is related to crash involvement. Therefore, it may be of particular benefit to older drivers if a salient signal is presented when a vehicle is in their blind spot and/or when they attempt to change lanes while another vehicle is in the intended path.

11.3 Adaptive cruise control

Adaptive cruise control (ACC) works much like standard cruise control but with the added advantage of also monitoring vehicle headway with a car in front. So, for example, if a driver sets the ACC at 55 mph and then encounters a vehicle in her lane traveling at 50 mph, the driver's car will slow to maintain a constant safe headway, rather than a constant speed. This feature has the potential to benefit all drivers.

11.4 Active lane keeping assistance

Active lane keeping (ALK) partially eliminates the need to steer by automatically centering the vehicle in the lane of travel. ALK may be of benefit to young and older drivers alike, particularly during high cognitive load situations, such as while searching for street signs or during heavy stop and go traffic. One key area that may be overlooked in the design is helping drivers develop an adequate mental model of the technology. In a recent study in our lab, three of six participants reported believing that ALK was capable of actively steering them into another lane to avoid obstacles in their current lane of travel. ALK is not designed to do this; therefore, this is a potentially serious misunderstanding of the technology. In fact, ALK has several known limitations, such as its reliance on lane demarcations and clear sensors. It is essential that instructions and guidance be designed to assist older adults in developing an adequate mental model of these systems.

11.5 Intersection and left-turn assist

ADASs that provide assistance with making left turns, particularly at intersections that are complex and/or that do not have left-turn signals can be particularly beneficial for older adults. Older drivers are overrepresented in crashes at intersections relative to the other age groups. Age-related changes in speed of processing, visual attention, and decreased physical mobility (discussed in Chapter 3) contribute to the difficulties older drivers have in making quick decisions about whether or not it is safe to turn across oncoming traffic when making a left turn.

One technology under development is stop sign movement assist. Stop sign movement assist is intended to assist drivers with making turns across traffic at intersections by warning of oncoming traffic and safe gap distances between cross-traffic cars. These systems are currently under development but do not yet exist in any vehicles on the market to the author's knowledge. Certain models of Volvo have "oncoming lane mitigation," which, in addition to braking, attempts to steer out of the path of an oncoming vehicle. Given the difficulties older adults have with negotiating intersections and judging safe gap distances (see discussion in Chapter 3), such systems could greatly benefit older adults. The systems will need to have clearly visible and understandable directions to avoid making an already challenging situation even more complex. Providing voice guidance in addition to a visual display will allow older adults to keep their eyes on the road and will reduce visual switch times (a problem for older adults discussed in Chapter 3).

A number of different systems have been developed; however, at present more research is needed to determine their overall benefit to older drivers. The existing evidence indicates that older drivers can benefit from an in-vehicle display that provides advance notice of upcoming intersections and the state of traffic signals, speed limits and gap sizes. These displays provide older drivers with more time to prepare for intersections and tend to decrease attention demands. Intersection and left-turn assist displays should include both a visual and auditory (or haptic) indicator to warn a driver when he or she is approaching a red light or if the driver attempts to make a left-hand turn when there is an insufficient gap (amount of time to make the turn based on the distance and speed of oncoming traffic) in the cross-traffic lane. These redundant (using two modalities) displays decrease visual processing demand and improve performance. Messages should be kept simple to avoid confusion or increased informational processing demands.

11.6 Design recommendations

- Provide adequate training on the capabilities and limitations of each ADAS to assist older drivers with developing an adequate mental model of the system—knowing situations in which the ADAS is likely to be helpful and its system limitations.
- Provide collision and blind-spot monitoring warnings in multiple modalities to offset age-related slowing in processing speed and to accommodate older adults who may have differential loss in sensory acuity in one or more modalities.

- Provide voice guidance to supplement visual displays to allow older drivers to keep their eyes on the road and minimize the need to switch between multiple visual displays.
- Intersection and left-turn assistance should use redundant coding (multiple modalities) and should provide only simple messages.

Recommended readings

Dotzauer, M. 2014. Researching safety issues with intersection assistance systems for the older driver (Chapter 7). In A. Stevens, C. Brusque, & J. Krems (Eds.), *Driver Adaptation to Information and Assistance Systems*. London, UK: Institution of Engineering and Technology.

McDonald, A., Carney, C., & McGehee, D. V. 2018. *Vehicle Owners' Experiences with and Reactions to Advanced Driver Assistance Systems*. https://aaafoundation.org/vehicle-owners-experiences-reactions-advanced-driver-assistance-systems/

Reimer, B. 2014. Driver assistance systems and the transition to automated vehicles: A path to increase older adult safety and mobility? *Public Policy and Aging Report*, 24(1), 27–31. doi: 10.1093/ppar/prt006.

chapter twelve

Summary, synthesis, and conclusions

The world's population is aging. Aging is accompanied by sensory, cognitive, and physical challenges that may compromise mobility. Yet, mobility is vital to maintaining health, social engagement, and quality of life. There is great interindividual variability among older adults, but most will experience some age-related changes that compromise the ease with which they can realize their transportation needs. Changes in transportation design that enhance the mobility of older adults can also enhance the experience and safety of all travelers. For example, advanced driver assistance systems (ADASs) and other form of automation have great potential for improving the safety of all drivers, with particular promise for maintaining the health, safety, and mobility of older drivers. As we have seen in the previous chapters, implementation of certain design guidelines can accomplish this goal. Key design guidelines and the overarching principles on which they are based are summarized in this final chapter. We see that reducing distraction, providing auditory assistance to offset high visual load, and ensuring that visual information is accessible to older adults are valuable steps to improving mobility.

12.1 Interface modalities guidelines

Sensory impairments are common in advanced age, with nearly 20% of adults over age 70 reporting significant visual impairment, nearly 33% reporting hearing impairment, and another 8% experiencing significant deficits in both visual and hearing abilities. Designing transportation interfaces that reduce the negative effect of these changes can have profound positive effects on the mobility and overall quality of life of older adults.

12.1.1 Auditory interfaces

Driving makes heavy demands on the visual system. Older adults often benefit from auditory interfaces that can offset the heavy visual load of driving and navigating complex areas. Older adults take more time to extract visual information and take longer to switch visual attention from one event or display to another. Therefore, providing auditory guidance (e.g., voice navigation, auditory menus, and reminders) can benefit older adults even more than their younger counterparts.

Older adults on average have experienced some hearing loss. To assist older listeners who have experienced age-related loss of auditory acuity, auditory information should be presented at least 10–15 dB above ambient background noise unless doing so poses a risk of greater hearing loss (e.g., keep overall noise levels below 85 dB). Further, providing a means of adjusting the sound to a comfortable listening level for each individual will increase the benefit of the auditory modality. This is particularly important for people with hyperacusis, or increased sensitivity to environmental noise for which age-related hearing loss is a risk factor. Allowing people to adjust the level of auditory alerts in their vehicles can mitigate that issue.

12.1.2 Visual interfaces

Age-related loss of accommodation due to hardening of the lens (presbyopia) makes it harder to read small font sizes even if distance vision is relatively unimpaired. Therefore, ensuring that font sizes are large, or can be adjusted, will aid the ease with which older adults can extract visual information and also will decrease the overall demand of the processing task. The lens also yellows with age, making it more difficult to distinguish between certain colors (e.g., blues and greens); therefore, it is best to avoid using these colors when possible and provide redundant coding (i.e., another means of differentiating like shape, size, or location). Those benefits can free resources for other aspects of the driving/navigation task.

12.1.3 Tactile and haptic interfaces

Vibrotactile and haptic interfaces are increasingly common. Cellular phones can be set to vibrate, and rumble strips are used to warn drivers when their car veers off the road. New technologies can simulate these road vibrations and use them as alerts presented through the seat pan or steering wheel. For example, General Motors has a "safety alert seat" that vibrates when a collision risk is detected. The use of vibrotactile interfaces (particularly in conjunction with redundant auditory or visual displays) can offset age-related slowed processing and/or response times, as discussed in Chapter 6.

12.1.4 Gesture-based interfaces

Gesture-based interfaces are increasingly making their way into modern automobiles. There is currently little research on how these interfaces will impact older adults. However, it can be expected that designs that allow for simple hand gestures and flexibility in precision will likely benefit older drivers who may find it difficult to use other forms of controls.

12.1.5 Multimodal interfaces

Using more than one modality to present critical information (e.g., such as collision warnings or pedestrian crossing information) increases the chances that older adults will perceive the signals. For example, an older adult with impaired hearing can still process the visual signal, while a person with visual impairment can still process the auditory signal. The use of multimodal interfaces (a form of redundant coding) can also speed processing time. Even young adults have been found to respond faster to a redundant visual and auditory signal than to a signal presented in only one modality.

Use of an auditory modality for time-critical signals is particularly important for older adults. Auditory signals capture attention regardless of the direction of gaze. Adding an auditory or tactile component to critical visual warnings, such as forward collision warnings and blind-spot warnings, can greatly improve the potential safety benefits for older adults. However, keep in mind that due to age-related increases in auditory hearing thresholds, the auditory component to critical warnings likely needs to be at least 10–15 dB louder on average in order to be heard by the older ear. It would be important for the sound level of the auditory component of the warning to be adjustable by the older individual to his or her threshold and comfort.

12.2 Reducing memory requirements

Remembering long strings of information, such as navigational directions, can be difficult at any age, but becomes particularly challenging for older adults who typically have some working memory deficits. To reduce working memory load, it is important to keep working memory requirements low by providing short, terse strings of information that can be easily rehearsed verbally by the driver.

12.2.1 Redundancy

Redundancy can improve information processing for older adults, as discussed earlier through the use of multimodal displays. Another form of redundant coding for displays and controls is the use of more than one dimension (even within a single modality) to aid in distinguishing between items. For example, controls can be distinguished by both color and shape, or size and location. Redundant visual coding can greatly aid persons with reduced vision. Common examples of redundant visual displays include the American stop sign and traffic signals at intersections. For the stop sign, both shape and color distinguish it from all other road signs. For the traffic signal, the red light signaling the need to stop is always located at the top, the green light is at the bottom, and the amber light is in between. The use of redundancy and consistency allows people who are color blind to still distinguish between the signals and speeds processing for all road users.

12.2.2 Repetition

Repetition, or at least the option to have verbal information or instructions repeated, is particularly helpful for older adults. It is also helpful for visual information when the information is presented in a transient form (does not stay visible on the display). For example, many navigation systems provide advance notice of upcoming turns and then repeat part or all of the information as the turn gets closer. Providing an option to request navigational instructions be repeated could also be helpful, though this is not currently available in many navigational aids. Providing both visual and verbal information supports repetition and is another form of redundancy that can aid older adults. Most navigational aids do present instructions in both map and narrative forms, though several key strokes could be required to switch between those. If the navigational instruction is missed, not clear, or forgotten, providing another iteration as well as presenting information both visually and verbally can improve the chances that important information is processed.

12.3 Multitasking and distraction

Driving, navigating through airports, and taking other forms of transportation require multitasking. Age-related cognitive changes decrease older adults' ability to multitask. At the same time, inhibitory deficits make older adults more prone to distraction from irrelevant information. Therefore, unnecessary distractions (e.g., cluttered road signs and advertisements) should be reduced or avoided.

12.3.1 Timing and prioritization

Age-related slowing of information processing means that older adults need advanced notification of turn information for navigational guidance systems and more time allowed for comprehending informational messages. A hierarchy of interface alerts is needed in order to avoid sending the driver multiple alerts in quick succession. The most critical alerts (e.g., collision warning) should be prioritized and sent first, with less critical alerts delayed in time (e.g., fatigue warning) in order not to interfere with processing.

12.4 General design recommendations

- Policy
  - Federal and state transportation agencies should take into account the physical and cognitive limitations of older people when forming policies. Policies should include alternatives to driving such as cycling and walking as well as health and safety.

- Environment outside vehicle and road infrastructure
  - Good lighting should be provided on roads, bridges, walkways, and in intersections.
  - Handrails should be available for pedestrians.
  - Intersections should be designed to avoid or improve safety for left turns (roundabouts, J-intersections, etc.).
  - Countdown displays should be added to crossing signals.
  - Pedestrian refuge islands, road diets, and pedestrian hybrid beacons should be used.
- Information outside vehicle in signs
  - Reduce information load.
  - Reduce memory requirements.
  - Increase redundancy.
  - Reduce display complexity, increase font size, and use light-colored characters on dark background.
  - Use retroreflective material.
  - Provide redundant information in work zones and where novel information is presented (e.g., airports).
  - Messages that involve more than one phase (message split in time in variable message signs) should appear in no more than two phases.
- Information inside vehicle
  - Reduce distractions, multitasking demands, and display complexity.
  - Provide auditory assistance to offset high visual load (auditory information recommendation is 10–15 dB above speech perception level).
  - Use a multimodal presentation of time-critical information (e.g., collision warnings).
  - Provide adjustable fonts for visual information.
  - Offer adjustable volume controls for navigation and alert sound levels.
  - Reduce glare on displays.
  - Reduce noise inside the vehicle to enhance "speech-in-noise" perception.
  - Develop gesture-based systems.
- Enhancing the vehicle
  - Offer automatic safety systems, including blind-spot detection, automatic emergency braking, and cross-traffic alerts.
  - Educate older adults about the ability of vehicle automatic safety systems to reduce crashes.
- Enhancing the driver
  - Encourage optimal optical correction.
  - Offer useful field of view training, intersection scanning training, and driving skills training.

- Improve speech-in-noise perception through auditory perception training.
- Participate in range-of-motion training to reduce crash risk.
- Participate in aerobic and strength training to benefit physical health and cognition.

Recommended readings

Fisk, A. D., Rogers, W. A., Charness, N., Czaja, S. J., & Sharit, J. 2009. *Designing for Older Adults: Principles and Creative Human Factors Approaches* (2nd ed.). Boca Raton, FL: CRC Press/Taylor and Francis Group.

Johnson, J., & Finn, K. 2017. *Designing User Interfaces for an Aging Population: Towards Universal Design*. Cambridge, MA: Morgan Kaufman/Elsevier.

References

Ahn, M. 2015, March 26. The MTA's Accessibility Gap. December 13, 2017. Retrieved from https://subwayrecord.wordpress.com/2015/03/26/the-mtas-accessibility-gap/

Anderson, S., White-Schwoch, T., Parbery-Clark, A., & Kraus, N. 2013. Reversal of age-related neural timing delays with training. *Proceedings of the National Academy of Sciences of the United States of America*, 110(11), 4357–4362. doi: 10.1073/pnas.1213555110

Anguera, J. A., Boccanfuso, J., Rintoul, J. L., Hashimi, O. A., Faraji, F., Janowich, J., ... Gazzaley, A. 2013. Video game training enhances cognitive control in older adults. (Research: Letter) (Report). *Bionature*, 501(7465), 97. doi: 10.1038/nature12486

Arch, A. 2010. Web accessibility for older users: A literature review. Retrieved from https://www.w3.org/TR/2008/WD-wai-age-literature-20080514/

Baddeley, A. D., & Hitch, G. 1974. Working memory. In G. H. Bower (Ed.), *The Psychology of Learning and Motivation* (Vol. 8, pp. 47–89). Orlando, FL: Academic Press.

Baldwin, G. 2010. *Aging, Transportation, and Health: U.S. Senate Special Committee on Aging*. Washington, DC: Department of Health and Human Services.

Ball, K., Beard, B. L., Roenker, D. L., Miller, R. L., & Griggs, D. 1988. Age and visual search: Expanding the useful field of view. *Journal of the Optical Society of America, A*, 5, 2210–2219.

Ball, K., Edwards, J. D., Ross, L. A., & McGwin, G., Jr. 2010. Cognitive training decreases motor vehicle collision involvement of older drivers. *Journal of the American Geriatrics Society*, 58(11), 2107–2113. doi: 10.1111/j.1532-5415.2010.03138.x

Bellet, T., Paris, J.-C., & Marin-Lamellet, C. 2018. Difficulties experienced by older drivers during their regular driving and their expectations towards Advanced Driving Aid Systems and vehicle automation. *Transportation Research. Part F, Traffic Psychology and Behaviour*, 52, 138.

Brown, K. 2003. *Staying Ahead of the Curve 2003: The AARP Working in Retirement Study*. Washington, DC: American Association of Retired Persons (AARP).

Burkhardt, J. E., Bernstein, D. J., Kulbicki, K., Eby, D. W., Molnar, L. J., Nelson, C. A., & McClary, J. M. 2014. Travel Training for Older Adults Part II: Research Report and Case Studies. In *Transit Cooperative Research Program* (Ed.), Washington, DC: Transportation Research Board of the National Academies.

Chevalier, A., Coxon, K., Chevalier, A. J., Clarke, E., Rogers, K., Brown, J., ... Keay, L. 2017. Predictors of older drivers' involvement in rapid deceleration events. *Accident Analysis and Prevention*, 98, 312–319. doi: 10.1016/j.aap.2016.10.010

Church, A., Frost, M., & Sullivan, K. 2000. Transport and social exclusion in London. *Transport Policy*, 7(3), 195–205. doi: 10.1016/S0967-070X(00)00024-X

Cicchino, J. B. 2017. Effectiveness of forward collision warning and autonomous emergency braking systems in reducing front-to-rear crash rates. *Accident Analysis and Prevention*, 99(Pt A), 142–152. doi: 10.1016/j.aap.2016.11.009

de Hartog, J., Boogaard, H., Nijland, H., & Hoek, G. 2010. Do the health benefits of cycling outweigh the risks? *Environmental Health Perspectives*, 118(8), 1109–1116. doi: 10.1289/ehp.0901747

Dingus, T. A., Guo, F., Lee, S., Antin, J. F., Perez, M., Buchanan-King, M., & Hankey, J. 2016. Driver crash risk factors and prevalence evaluation using naturalistic driving data. *Proceedings of the National Academy of Sciences of the United States of America*, 113(10), 2636–2641. doi: 10.1073/pnas.1513271113

Dumbaugh, E., & Zhang, Y. 2013. The relationship between community design and crashes involving older drivers and pedestrians. *Journal of Planning Education and Research*, 33(1), 83–95. doi: 10.1177/0739456X12468771

Elosua, P. 2011. Subjective values of quality of life dimensions in elderly people. A SEM preference model approach. *Social Indicators Research*, 104(3), 427–437. doi: 10.1007/s11205-010-9752-y

Evans, D. W., & Ginsburg, A. P. 1985. Contrast sensitivity predicts age-related differences in highway-sign discriminability. *Human Factors*, 27(6), 637–642. doi: 10.1177/001872088502700602

Farquhar, M. 1995. Elderly people's definitions of quality of life. *Social Science and Medicine*, 41(10), 1439–1446. doi: 10.1016/0277-9536(95)00117-P

Federal Highway Administration (FHWA). 2014. *Median U-Turn Informational Guide*. (FHWA-SA-14-069). Washington, DC: US Department of Transportation. Retrieved from https://safety.fhwa.dot.gov/intersection/alter_design/pdf/fhwasa14069_mut_infoguide.pdf

Folstein, M. F., Folstein, S. E., & McHugh, P. R. 1975. "Mini-Mental State": A practical method for grading the cognitive state of patients for the clinician. *Journal of Psychiatric Research*, 12(3), 189–198. doi: 10.1016/0022-3956(75)90026-6

Green, K. A., McGwin, G., Jr., & Owsley, C. 2013. Associations between visual, hearing, and dual sensory impairments and history of motor vehicle collision involvement of older drivers. *Journal of the American Geriatrics Society*, 61(2), 252–257. doi: 10.1111/jgs.12091

Greenwood, P. M., & Parasuraman, R. 2012. *Nurturing the Older Brain and Mind*. Cambridge, MA: The MIT Press.

Greenwood, P. M., & Parasuraman, R. 2016. The mechanisms of far transfer from cognitive training: Review and hypothesis. *Cognitive Neuropsychology*, 30(6), 742–755. doi: 10.1037/neu0000235

Hickman, J. M., Rogers, W. A., & Fisk, A. D. 2007. Training older adults to use new technology. *Journals of Gerontology. Series B, Psychological Sciences and Social Sciences*, 62(Spec No 1), 77–84.

Horswill, M. S., Falconer, E. K., Pachana, N. A., Wetton, M., & Hill, A. 2015. The longer-term effects of a brief hazard perception training intervention in older drivers. *Psychology and Aging*, 30(1), 62–67. doi: 10.1037/a0038671

Keay, L., Jasti, S., Munoz, B., Turano, K. A., Munro, C. A., Duncan, D. D., … West, S. K. 2009. Urban and rural differences in older drivers' failure to stop at stop signs. *Accident Analysis and Prevention*, 41(5), 995–1000. doi: 10.1016/j.aap.2009.06.004

Kim, H., Kwon, S., Heo, J., Lee, H., & Chung, M. K. 2014. The effect of touch-key size on the usability of in-vehicle information systems and driving safety during simulated driving. *Applied Ergonomics*, 45, 379–388. doi: 10.1016/j.apergo.2013.05.006

Koepsell, T., McCloskey, L., Wolf, M. et al. 2002. Crosswalk markings and the risk of pedestrian–motor vehicle collisions in older pedestrians. *JAMA*, 288(17), 2136–2143. doi: 10.1001/jama.288.17.2136

Langford, J., Charlton, J. L., Koppel, S., Myers, A., Tuokko, H., Marshall, S., ... Macdonald, W. 2013. Findings from the Candrive/Ozcandrive study: Low mileage older drivers, crash risk and reduced fitness to drive. *Accident Analysis and Prevention*, 61, 304–310. doi: 10.1016/j.aap.2013.02.006

Laurienti, P. J., Burdette, J. H., Maldjian, J. A., & Wallace, M. T. 2006. Enhanced multisensory integration in older adults. *Neurobiology of Aging*, 27(8), 1155–1163. doi: 10.1016/j.neurobiolaging.2005.05.024

Lewis, B. A., Eisert, J. L., Baldwin, C. L., Singer, J., & Lerner, N. 2017, March. *Urgency Coding Validations (National Highway Traffic Safety Administration)*. Washington, DC: US Department of Transportation.

Li, R. C., Yingjie, V., Sha, C., & Lu, Z. 2017. Effects of interface layout on the usability of in-vehicle information systems and driving safety. *Chinese Journal of Liquid Crystals and Displays*, 49, 124–132. doi: 10.1016/j.displa.2017.07.008

Marcum, C. S. 2013. Age differences in daily social activities. *Research on Aging*, 35(5), 612–640. doi: 10.1177/0164027512453468

Margrain, T. H., & Boulton, M. 2005. Sensory impairment. In M. L. Johnson (Ed.), *The Cambridge Handbook of Age and Ageing* (pp. 121–129). Cambridge, UK: Cambridge University Press.

Marottoli, R., Allore, H., Araujo, K., Iannone, L., Acampora, D., Gottschalk, M., ... Peduzzi, P. 2007a. A randomized trial of a physical conditioning program to enhance the driving performance of older persons. *Journal of General Internal Medicine*, 22(5), 590–597. doi: 10.1007/s11606-007-0134-3

Marottoli, R. A., Cooney, L. M., Wagner, R., Doucette, J., & Tinetti, M. E. 1994. Predictors of automobile crashes and moving violations among elderly drivers. *Annals of Internal Medicine*, 121(11), 842. doi: 10.7326/0003-4819-121-11-199412010-00003

Marottoli, R. A., Van Ness, P. H., Araujo, K. L. B., Iannone, L. P., Acampora, D., Charpentier, P., & Peduzzi, P. 2007b. A randomized trial of an education program to enhance older driver performance. *Journals of Gerontology, Series A, Biological Sciences and Medical Sciences*, 62, 1113–1119.

Mielenz, T., Durbin, L., Cisewski, J., Guralnik, J., & Li, G. 2017. Select physical performance measures and driving outcomes in older adults. *Injury Epidemiology*, 4(1), 1–15. doi: 10.1186/s40621-017-0110-2.

Mitolo, M., Borella, E., Meneghetti, C., Carbone, E., & Pazzaglia, F. 2016. How to enhance route learning and visuo-spatial working memory in aging: A training for residential care home residents. *Aging and Mental Health*, 1–9. doi: 10.1080/13607863.2015.1132673

Munro, C. A., Jefferys, J., Gower, E. W., Munoz, B. E., Lyketsos, C. G., Keay, L., ... West, S. K. 2010. Predictors of lane-change errors in older drivers. *Journal of the American Geriatrics Society*, 58(3), 457–464. doi: 10.1111/j.1532-5415.2010.02729.x

National Institute on Aging (NIA). 2004. *Older Drivers*. Bethesda, MD: NIA, US Department of Health and Human Services, Public Health Service, National Institutes of Health.

National Safety Council. 2017. *2017 Fatality Estimates*. Retrieved from https://www. nsc.org/road-safety/safety-topics/fatality-estimates

NHTSA, Traffic Safety Facts. 2016. *NHTSA Traffic Safety Facts 2016 812456*.

Norton, S., Matthews, F. E., Barnes, D. E., Yaffe, K., & Brayne, C. 2014. Potential for primary prevention of Alzheimer's disease: An analysis of population-based data. *Lancet Neurology*, 13(8), 788–794. doi: 10.1016/S1474-4422(14)70136-X

Ostrow, A. C., Shaffron, P., & McPherson, K. 1992. The effects of a joint range-of-motion physical fitness training program on the automobile driving skills of older adults. *Journal of Safety Research*, 23(4), 207–219. doi: 10.1016/0022-4375(92)90003-R

Paire-Ficout, L., Marin-Lamellet, C., Lafont, S., Thomas-Anterion, C., & Laurent, B. 2016. The role of navigation instruction at intersections for older drivers and those with early Alzheimer's disease. *Accident Analysis and Prevention*, 96, 249–254. doi: 10.1016/j.aap.2016.08.013

Pollatsek, A., Romoser, M. R. E., & Fisher, D. L. 2012. Identifying and remediating failures of selective attention in older drivers. *Current Directions in Psychological Science*, 21(1), 3–7. doi: 10.1177/0963721411429459

Reagan, I. J., Kidd, D. G., Dobres, J., Mehler, B., & Reimer, B. 2017. The effects of age, interface modality, and system design on drivers' attentional demand when making phone calls while driving on a limited-access highway. Report authored by the Insurance Institute for Highway Safety (IIHS).

Retting, R. A., Ferguson, S. A., & McCartt, A. T. 2003. A review of evidence-based traffic engineering measures designed to reduce pedestrian-motor vehicle crashes. *American Journal of Public Health*, 93(9), 1456–1463.

Risacher, S. L., McDonald, B. C., Tallman, E. F., West, J. D., Farlow, M. R., Unverzagt, F. W., … Alzheimer's Disease Neuroimaging, Initiative. 2016. Association between anticholinergic medication use and cognition, brain metabolism, and brain atrophy in cognitively normal older adults. *JAMA Neurology*, 73(6), 721–732. doi: 10.1001/jamaneurol.2016.0580

Rizzo, M. 2011. Impaired driving from medical conditions: A 70-year-old man trying to decide if he should continue driving. *JAMA*, 305(10), 1018–1026. doi: 10.1001/jama.2011.252

Roenker, D. L., Cissell, G. M., Ball, K. K., Wadley, V. G., & Edwards, J. D. 2003. Speed-of-processing and driving simulator training result in improved driving performance. *Human Factors*, 45(2), 218–233. doi: 10.1518/hfes.45.2.218.27241

Ross, L. A., Edwards, J. D., O'Connor, M. L., Ball, K. K., Wadley, V. G., & Vance, D. E. 2016. The transfer of cognitive speed of processing training to older adults' driving mobility across 5 years. *Journals of Gerontology. Series B, Psychological Sciences and Social Sciences*, 71(1), 87–97. doi: 10.1093/geronb/gbv022

Rouxel, P., Webb, E., & Chandola, T. 2017. Does public transport use prevent declines in walking speed among older adults living in England? A prospective cohort study. *BMJ Open*, 7(9), e017702. doi: 10.1136/bmjopen-2017-017702

Sawula, E., Polgar, J., Porter, M. M., Gagnon, S., Weaver, B., Nakagawa, S., … Bédard, M. 2017. The combined effects of on-road and simulator training with feedback on older drivers' on-road performance: Evidence from a randomized-controlled trial. *Traffic Injury Prevention*, 19(3), 241–249. doi: 10.1080/15389588.2016.1236194

Sklar, A. L., Boissoneault, J., Fillmore, M. T., & Nixon, S. J. 2014. Interactions between age and moderate alcohol effects on simulated driving performance. *Psychopharmacology*, 231(3), 557–566. doi: 10.1007/s00213-013-3269-4

Smith, P., Blumenthal, J., Hoffman, B., Cooper, H., Strauman, T., Welsh-Bohmer, K., … Sherwood, A. 2010. *Aerobic Exercise and Neurocognitive Performance: A Meta-Analytic Review of Randomized Controlled Trials* (Vol. 72, pp. 239). Baltimore, MD: Lippincott Williams & Wilkins, Ovid Technologies.

Strayer, D. L., Cooper, J. M., McCarty, M. M., Getty, D. J., Wheatley, C. L., Motzkus, C. J., … Biondi, F. 2018. *Visual and Cognitive Demands of Using Apple's CarPlay, Google's Android Auto and Five Different OEM Infotainment Systems.* AAA Foundation for Traffic Safety.

Transport Council. 1998. *SARA: The Top Four Issues.* Clerkenwell Green, London, UK: Help the Aged Transport Council.

U.S. Department of Transportation. 2017. Federal Highway Administration, 2017. *National Household Travel Survey.* Retrieved from http://nhts.ornl.gov

Valenti, G., Bonomi, A. G., & Westerterp, K. R. 2016. Multicomponent fitness training improves walking economy in older adults. *Medicine and Science in Sports and Exercise*, 48(7), 1365. doi: 10.1249/MSS.0000000000000893

Wilhelmson, K., Andersson, C., Waern, M., & Allebeck, P. 2005. Elderly peoples' perspectives on quality of life. *Ageing and Society*, 25(4), 585–600. doi: 10.1017/S0144686X05003454

Wood, J. M., & Owens, D. A. 2005. Standard measures of visual acuity do not predict drivers' recognition performance under day or night conditions. *Optometry and Vision Science*, 82(8), 698–705.

Zwald, M. L., Hipp, J. A., Corseuil, M. W., & Dodson, E. A. 2014. Correlates of walking for transportation and use of public transportation among adults in St Louis, Missouri, 2012. *Preventing Chronic Disease*, 11, E112. doi: 10.5888/pcd11.140125

Index